INVESTIGATING HUMAN DISEASES WITH THE MICROBIOME

Metagenomics Bench to Bedside

知菌知病

—— 宏基因组精准健康的未来

贾慧珏（Huijue Jia） 著

贾慧珏（Huijue Jia） 李莹（Ying Li） 主译

清华大学出版社
北 京

北京市版权局著作权合同登记号　图字：01-2022-6685

Investigating Human Diseases with The Microbiome: Metagenomics Bench to Bedside
Huijue Jia
ISBN: 978-0-323-91369-0
Copyright © 2022 Elsevier Inc. All rights reserved.
Authorized Chinese translation published by Tsinghua University Press.

《知菌知病——宏基因组精准健康的未来》（贾慧珏，李莹主译）
ISBN: 978-7-302-67503-7
Copyright © Elsevier Inc. and Tsinghua University Press. All rights reserved.

No part of this publication may be reproduced or transmitted in any form or by any means, electronic or mechanical, including photocopying, recording, or any information storage and retrieval system, without permission in writing from Elsevier (Singapore) Pte Ltd. Details on how to seek permission, further information about the Elsevier's permissions policies and arrangements with organizations such as the Copyright Clearance Center and the Copyright Licensing Agency, can be found at our website: www.elsevier.com/permissions.

This book and the individual contributions contained in it are protected under copyright by Elsevier Inc. and Tsinghua University Press (other than as may be noted herein).

This edition of Investigating Human Diseases with The Microbiome: Metagenomics Bench to Bedside is published by Tsinghua University Press under arrangement with ELSEVIER INC.

This edition is authorized for sale in the People's Republic of China only, excluding Hong Kong, Macao SAR and Taiwan. Unauthorized export of this edition is a violation of the Copyright Act. Violation of this Law is subject to Civil and Criminal Penalties.

本版由ELSEVIER INC.授权清华大学出版社出版。此版本仅限在中华人民共和国境内（不包括中国香港、澳门特别行政区和台湾地区）销售。未经许可之出口，视为违反著作权法，将受民事及刑事法律之制裁。
本书封底贴有Elsevier防伪标签，无标签者不得销售。

声　明

本书涉及领域的知识和实践标准在不断变化。新的研究和经验拓展我们的理解，因此须对研究方法、专业实践或医疗方法作出调整。从业者和研究人员必须始终依靠自身经验和知识来评估和使用本书中提到的所有信息、方法、化合物或本书中描述的实验。在使用这些信息或方法时，他们应注意自身和他人的安全，包括注意他们负有专业责任的当事人的安全。在法律允许的最大范围内，爱思唯尔、译文的原文作者、原文编辑及原文内容提供者均不对因产品责任、疏忽或其他人身或财产伤害及/或损失承担责任，亦不对由于使用或操作文中提到的方法、产品、说明或思想而导致的人身或财产伤害及/或损失承担责任。

本书封面贴有清华大学出版社防伪标签，无标签者不得销售。
版权所有，侵权必究。举报：010-62782989，beiqinquan@tup.tsinghua.edu.cn。

图书在版编目（CIP）数据

知菌知病：宏基因组精准健康的未来 / 贾慧珏著；贾慧珏，李莹主译. -- 北京：清华大学出版社，2024.10. -- ISBN 978-7-302-67503-7

Ⅰ. Q343.2
中国国家版本馆 CIP 数据核字第 2024M3X847 号

责任编辑：辛瑞瑞　孙　宇
封面设计：钟　达
责任校对：李建庄
责任印制：曹婉颖

出版发行：清华大学出版社
　　　　　网　　址：https://www.tup.com.cn, https://www.wqxuetang.com
　　　　　地　　址：北京清华大学学研大厦 A 座　　邮　编：100084
　　　　　社 总 机：010-83470000　　　　　　　　邮　购：010-62786544
　　　　　投稿与读者服务：010-62776969，c-service@tup.tsinghua.edu.cn
　　　　　质量反馈：010-62772015，zhiliang@tup.tsinghua.edu.cn
印 装 者：三河市龙大印装有限公司
经　　销：全国新华书店
开　　本：185mm×260mm　　　印　张：13　　　字　数：256 千字
版　　次：2024 年 12 月第 1 版　　　　　　　　印　次：2024 年 12 月第 1 次印刷
定　　价：128.00 元

产品编号：098235-01

译者名单

主　译　贾慧珏　李　莹
副主译　郑　晓

序　言

自人类基因组计划（Human Genome Project，HGP）保质、保量、提前两年于2003年4月宣布完成既定计划以来，以基因组学为先导的"组学"（omics）已经有了新的、重大的进展，并且催生了很多新的学科，其中最为重要的新学科便是宏基因组学（metagenomics）。

宏基因组学（尽管中译词在学术方面仍有不同意见，另一中译是"微生物组群"）最为重要的意义是使我们第一次较为充分地意识到人类还有"另外一个基因组"（our other genome），这使我们对微生物世界以及整个生命与非生命世界的相互作用方面的理解上升到了一个新的高度。

宏基因组学的研究一般是指一个环境或自然（生态）的样本中的所有微生物物种的全基因组分析研究，这也在一定意义上丰富了我们对生物"样本"的认识。当然，宏基因组也针对我们人体器官及体液中的所有微生物，包括细菌、病毒以及真菌，这些微生物组群存在于我们的胃肠道及所有其他组织/器官及体液之中，与食物消化、营养摄入、物质/能量代谢、废物排泄、疾病发生，以至于信号传导及基因调控网络息息相关。

自HGP完成以来，宏基因组学得到了国内外研究者的广泛关注，特别是中国科学家于2010年在《自然》杂志上发表第一张关于人类肠道微生物组群基因的综合目录，在癌症和其他疾病中都发现了很多关联微生物，可以说这是人类对自己了解的一小步，也是我们对体内以至于整个生命世界多样性的认识的一大步。

从技术方面来说，宏基因组学技术摒弃了传统上的微生物研究总是先分离、再培养至纯种鉴定，最后再测序进行分析的传统程序（即摸索培养条件并分别培养分离出每种微生物等烦琐步骤），而是对所研究的固体或液体样本直接"先测序后解序"，以了解样品所含的所有微生物的群体大小以及群体之间的各种互作关系。这一技术目前已经推广至人体以外的整个生态环境，例如运用宏基因组学技术对体外微生物（病原、病原宿主）的检测，在新冠肺炎疫情中已经做出了非常大的贡献。全球多个国家和地区已经利用宏基因组学技术在下水道等地区开展疫情防控，可以精准、及时地掌握一个居民小区乃至于一个小城的病毒流行及感染情况。

本书的作者多年来一直进行各种宏基因组学的研究。她和她的团队志在突破、勇于创新，并广泛参与国际合作，在这个领域很有建树。本书作者以其丰富的实践经验和扎实的文学功底，把一个新的学科、高深莫测的科学知识和新的学术名词以通俗易懂的语言来表述，这既可以帮助专业人士进一步学习该领域的科学知识和新近研究进展，也是对广大群众最好的科普。

我十分乐意为本书作序，并真诚希望此书能激励吾辈与晚辈对微生物组群的探索和更为深入的研究，并衷心希望作者和她的研究团队"百尺竿头，更进一步"，能够更好地应对全球人类群体层面的各种疾病防治及生物多样性的保护，敬畏生命，了解生命，探索自然规律而更好地造福人类。

杨焕明

中国科学院院士

中国医学科学院学部委员

发展中国家科学院（TWAS）成员

欧洲分子生物学组织（EMBO）成员

美国、德国及丹麦等国科学院的国际/外籍院士

译者序

从意识到自身肉体之躯蕴藏着数以十万亿计微小生命的那一天起，人类就没有停止过对它们的探索。17世纪的欧洲经济重镇荷兰代尔夫特，孕育出了列文虎克这样醉心科研的跨界人才，拉开了微生物研究序幕，300多年后的21世纪，我们终于迎来新技术的重大突破——低成本的高通量测序，这才帮助人体共生微生物研究正式迈入高速发展赛道。

越来越多的研究成果表明，对人体自身及外环境的研究无法完全解释疾病发生发展的机制，解开这个难题的关键性钥匙很可能藏在人体共生微生物与健康的探索研究当中。作为研究共生微生物最重要的创新方法，以宏基因组学为代表的多种组学新兴学科均具备多领域交叉融合的典型特征。开展组学研究必须面对的海量碎片化数据的获取、处理与挖掘，对传统研究人员来说无疑困难重重。科研工作开始之初，研究者通常要面对一系列宏基因组学相关的通识性问题，如什么是微生物群落与微生物组？如何科学高效地收集宏基因组学样本？不同部位人体共生微生物种群有什么特点？病原菌与共生微生物之间的界限又在哪里？为了解答这些问题，一本化繁为简再以简驭繁的研究指导应运而生。本书原作者贾慧珏博士作为共生微生物领域国际知名科学家，有感于跨领域研究者们与新兴学科初接触时的疑惑与困扰，将10年来前沿实践经历心得提炼归纳，集结成书，希望能帮助到打算开展人体共生微生物研究的人员，对其科研思路从无序到有序的转变过程进行引导。

本书总结了从微生物组角度研究人类健康和疾病的原理。在人的一生中，粪便、口腔、鼻腔、阴道和皮肤样本中的微生物组包含可以预测未来疾病风险的重要信息。组织样本中也含有与疾病相关的微生物。微生物组连接了遗传和环境因素，有望极大地促进精准医疗，包括预防、诊断和有效治疗许多复杂疾病。

本书基于传统微生物学，并结合高通量测序的更全面的视角，提出了关于人体内微生物细胞总数及其分布的关键问题，同时考虑了宏观生态学和因果推理的概念。书中专门介绍了各种方法，提供了关于收集宏基因组学研究样本时重要考虑事项的实践信息。

本书的重要特色在于提供了基于超过10年人类微生物组研究的框架，用于研究不同身体部位的微生物组为希望了解与人类健康相关的微生物的读者整合了相关信息,阐明了为什么可能希望在当前或未来的研究努力中包括微生物组研究提供了设计和进行微生物组研究和应用的技术考虑。

本书的撰写初衷是为广大医学生与医学工作者提供一本入门指导工具书,读者可以从中找到新手需要关注的所有基本要点。值得一提的是,全书提纲框架打破了按人体部位分类介绍的常规模式,开创性地从开展一项全新研究的角度,沿着研究方案设计的思维脉络徐徐展开。传播信息的同时,还分享了成熟科研工作者应该具备的思考问题的方式。作为一名宏基因组成果转化从业人员,译者在深入精读全书的过程中感受到了拓宽思路与精准提问带来的精神快感,而书中关于共生微生物间条件性互助或竞争关系的介绍充满了生物本能的智慧,给译者的工作乃至生活都带来了新的启示。为了把这样一本内容翔实且深入浅出的科普向书籍更好地介绍给大家,译者在翻译的过程中自行补充了许多解释说明,希望能为无论是来自泛理工科,还是较少接触专业工具书的潜在读者提供满意的阅读体验。

《宏基因组的精准医学未来》为译者参与完成的第一本宏基因组学指导书籍,虽然书中没有涉及主要为工具性质的计算机语言与程序编写相关教程介绍,却无法绕过数学与统计学的方法描述直接展示结论。在翻译的过程中,译者尽了最大努力去请教专业人员学习解释相关术语,如有错漏,敬请广大读者批评指正。

原版措辞审慎,逻辑清晰,用一种极简而不失趣味的方式,条理分明地介绍了共生微生物研究冗长纷杂的发展历史,并将爆发式增长的研究成果进行了总结升华。非常推荐对本书感兴趣的大家阅读原版,感受原作遣词造句中的严谨与简约之美。

译者 于深圳
2022 年 7 月

目 录

第1章 超有机体——微生物与宿主共生 ························· 1
    1.1 溯古追源——新技术带来的新发现 ·················· 1
    1.2 人体内有多少微生物细胞? ························· 8
    1.3 人体中的病毒微粒 ······························· 14
    1.4 其他物种中的微生物组 ··························· 15
    1.5 来自远古的微生物组 ····························· 15
    1.6 总结 ··· 18

第2章 微生物群落生态 ···································· 21
    2.1 宏观生态学中的营养级 ··························· 21
    2.2 微生物群落稳定性、多样性及丰富度 ················ 22
    2.3 皮肤微生物的几种不同生境 ······················· 27
    2.4 影响口腔微生物的因素 ··························· 29
    2.5 一个稳定的肠道菌群 ····························· 32
    2.6 肠型和塞伦盖蒂法则 ····························· 38
    2.7 总结 ··· 41

第3章 宏基因组样本收集 ·································· 55
    3.1 样本中非微生物的部分,会影响DNA提取和测序量 ······ 55
    3.2 对于粪便及生物量较低的宏基因组,要注意减少采样过程中
        每一步的污染 ·································· 57
    3.3 采样后,防止微生物增殖的试剂 ···················· 62
    3.4 对于宏基因组样本的DNA提取方法 ················· 63
    3.5 测序量 ······································· 64
    3.6 分类和功能概况,以及绝对丰度 ···················· 65
    3.7 宏基因组关联分析的样本量 ······················· 66
    3.8 总结 ··· 69

第4章 人体中的流行病学 ·································· 79
    4.1 和新型冠状病毒疫情类比 ························· 79

- 4.2 婴儿肠道共生菌的来源 ……………………………………………… 81
- 4.3 共生微生物的异位存续 ……………………………………………… 82
- 4.4 病灶部位菌群 ………………………………………………………… 82
- 4.5 微生物组中的跨界相互作用 ………………………………………… 89
- 4.6 其他显示微生物存在差异的组学数据 ……………………………… 90
- 4.7 总结 …………………………………………………………………… 94

第 5 章 演化中的微生物分类 …………………………………………… 102
- 5.1 对于常规应用的固定参考集 ………………………………………… 102
- 5.2 随着分类单元精度提升，数据变得稀疏 …………………………… 110
- 5.3 物种水平以下的微生物演化历程 …………………………………… 112
- 5.4 全细胞模型以实现从基因组到功能的完整预测 …………………… 116
- 5.5 总结 …………………………………………………………………… 118

第 6 章 共生微生物的疾病因果关系 …………………………………… 124
- 6.1 因果推断 ……………………………………………………………… 124
- 6.2 人类微生物和疾病之间关系的当前证据等级 ……………………… 129
- 6.3 从微生物到分子 ……………………………………………………… 132
- 6.4 总结 …………………………………………………………………… 137

第 7 章 宏基因组的临床应用 …………………………………………… 145
- 7.1 疾病筛查领域的宏基因组研究 ……………………………………… 145
- 7.2 临床实践启发深入研究 ……………………………………………… 157
- 7.3 利用微生物组知识重新定义现有疾病分类的可能性 ……………… 159
- 7.4 总结 …………………………………………………………………… 161

第 8 章 微生物档案描绘生命轨迹 ……………………………………… 173
- 8.1 在重要阶段进行前瞻性微生物采样 ………………………………… 173
- 8.2 从基因风险到疾病预防 ……………………………………………… 180
- 8.3 总结 …………………………………………………………………… 182

致　谢 ……………………………………………………………………… 195

第1章

超有机体——微生物与宿主共生

摘　要：从列文虎克观察牙菌斑中的微生物开始，本章节沿着微生物学中的技术进步的脚步，介绍了针对人体微生物的宏基因组研究。随着共生动物体型的增大，其体内微生物的个数会增多。人类结肠中微生物细胞的数量最多，为 $3.8×10^{13}$ 个，在结肠以外的黏膜表面、组织内部或体液中，数量相对较少。"宏基因组"（metagenome）、"微生物群落"（microbiota）和"微生物组"（microbiome）这三个名词已经存在了几十年，其中"微生物组"的内涵最广泛。

关键词：显微镜，微生物学，微生物短序列测序，微生物组，超级生物，共生功能体，微生物数量

1.1 溯古追源——新技术带来的新发现

列文虎克热衷于用他的"神秘工具"去观察一切他能想到的事物，这个"神秘工具"是一台单式显微镜，它的分辨率可以达到（在那个年代）令人吃惊的 1 μm（10^{-6} m，图1.1）[1,2]。尽管与他同时代的人相比，列文虎克已经非常注重口腔卫生与清洁，但他还是从包括他自己在内的几个人的牙菌斑中，吃惊地观察到了几种不同的细菌并对它们进行了描述（示例1.1）[3]。关于这些细菌的大小对比，列文虎克提到了沙粒。（示例1.1）沙粒直径一般是亚毫米级别，而比较小的那些沙粒直径在 100 ~ 200 μm，还在人眼的分辨范围内。

在胡克（R. Hooke）用他分辨率不高但更易于使用的双镜头显微镜（复式显微镜）[1]重现了列文虎克的"胡椒水实验"之后，列文虎克的各种研究获得了更多的认可。尽管如此，1758年林奈（C. Linnaeus）在编撰第10版《自然系统》（*Systema Naturae*）时仍然没有给微生物留出一个位置，他绘制的生命之树只包含植物和动物。后来生命之树的结构层次随着时代的发展发生了几次演化变迁。1866年 Ernest Haeckel 引入了"protista"（单细胞生物）这个概念。而名词"Monera"最初被用来描述缺少细胞核的单细胞生物，比如细菌。这个分类最初被归类为一个门，随后被提升为了一个单独的界。1969年，Robert Whittaker 为生命之树引入了真菌界，之后 Carl Woese 根据16S核糖体RNA序列比较，改进了生命之树，加入了古菌域[4]。

示例 1.1　列文虎克的牙齿细菌观察记，1683

虽然唾液在某种程度上不含有“animalcule”（列文虎克将明显可以移动的微小生物称之为 animalcule），但是正如他在 1683 年 9 月 17 日写给伦敦皇家学会的信中描述的那样，列文虎克在关于牙菌斑中的细菌的研究工作中，充分展示了他在科学方面的严谨性。

1）列文虎克的口腔清洁方式，以及他对自身牙菌斑样本的数次观察

“我习惯每天早上用盐清洁牙齿，然后用水漱口。每次吃过东西，我通常都会用牙签清理我的臼齿，并且用一块布用力擦拭牙齿。这使得我的牙齿始终保持清洁和洁白，在我这个年龄层的人中，很少有可以与我相提并论的。此外，当我用粗盐粒摩擦牙龈时，它们也不会出血。然而，当我用放大镜来观察牙齿时，这一切举措并没有让我的牙齿变得比其他人更干净，我可以看到有些东西黏在或长在臼齿和其他牙齿的缝隙里。这些东西是一些白色物质，像面糊一样。通过观察，我判断，虽然我看不到这里有任何东西在移动，但这些物质里面应该有会活动的小生物。于是我将这些牙缝间的白色物质与没有微生物的纯净雨水，以及从我的嘴里吸出的消除气泡后的唾液混合了几次（消除气泡是为了避免这些气泡搅动唾液）。我惊讶地发现，这些白色物质中有许多小小的、活生生的微小生物，它们移动的样子非常动人。在这些微小生物中，相对比较大的那些微生物的形状（图 1.1A），它们的动作非常迅速而有力，像梭子鱼一样在水或唾液中游动，数量并不算多。第二类微生物形状如图 1.1B 所示，它们经常像陀螺一样旋转，并且时不时像图 1.1C 到图 1.1D 中展示的那样运动，数量相对多一些。而第三类微生物，我看不出它们是什么形状，因为有时候看起来既有长的也有圆的，而另一些时候又都是圆的。它们是如此的小，我从未见到过它们比图 1.1E 所示更大的样子。它们飞快地向前移动，彼此之间密集地盘旋在一起，很容易让人联想到一大群蚊蚋或苍蝇聚成一团飞来飞去。这些时不时出现的小东西在我看

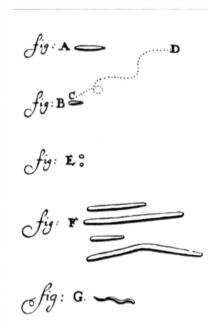

fig: A
D
fig: B C
fig: E
fig: F
fig: G.
Fig 2 ■ Engraving of oral bacteria by Abraham de Blois, from Leeuwenhoek's *Opera Omnia (Arcana Naturae Detecta)*, vol 2, Langerak, Arnold, 1722, p 40.

图 1.1　列文虎克观察到的微小生物

来是如此之多，以至于我估计在一滴与沙粒差不多大小的水或唾液（混合了前述提到过的白色物质）中，看到了几千个这样的东西，尽管这些从我的门牙和磨牙间取出的物质有9/10都是水或唾液。还有一些其他物质，大部分由大量的纤维状物质组成的，其中有些物质在长度方面格外与众不同，但它们在粗细程度上是一致的。它们如图1.1F所示的那样有些弯曲着，有些很直，乱糟糟地混合在一起。由于我以前在水里观察到过形状相同的活的微生物，所以我竭尽所能地去探索它们是否存在生命相关特质，但是最终我也没能从它们中观察到任何一点看起来类似具有生命的活动。"

2）对其他不同性别、年龄、口腔卫生习惯及吸烟/喝酒习惯的人群口腔样本进行观察

"我还从两位女性的嘴里收集了一些唾液，我确信这两位女性每天都会清洁口腔。我竭尽所能地检查了这些唾液，但是并没有从这里面识别到任何活着的微小生物。随后，我将同样的唾液与我用针从这两位女性的牙齿之间挑出的少量物质混合，果然发现了数量众多的活体微小生物和长微粒，与之前观察到的完全相同。"

"我还检查了一个大约8岁的孩子的唾液，但也没有发现任何活的微生物；在那之后，我把唾液和从这个孩子牙齿间取出的一些物质混合在一起，发现了大量之前提到过的微小生物和其他微粒。"

"一次，一位生活有节制、从不喝酒、从不抽烟的老人与我谈话，当我的视线落在他的牙齿上时，发现那里覆盖着一层薄膜。于是我询问他上一次清洁口腔是在什么时候，得到的回答是，他这一生从来都没有清洁过自己的口腔。后来我从他的嘴里取了一些唾液并检查，没能从中找到任何东西。"

"我还取了卡在他牙缝中间的那些物质，然后与没有微生物的清水混匀，同样也分了一部分与他的唾液混匀，然后我观察到了数量惊人的活体微小生物，它们游动的样子是如此灵活，远超我此前所见。体积较大的一个类别数量非常丰富，它们在前进的过程中会把身体弯曲成如图1.1G所示的曲线。此外，其他微小生物的数量众多，以至于整滴水都仿佛活了过来，尽管我只是从牙缝中取出了非常微量的物质与水混合。"

"我还取了另外一位老人的唾液以及黏附在他牙齿上牙缝中的白色物质。这位老人有上午喝烧酒、下午喝葡萄酒抽烟的习惯，我想知道这些小生物能否在这样持续饮酒的环境中存活下来。鉴于他的牙齿脏到匪夷所思，我断定此人从不清洁口腔。果然当我询问他时，他回答道：'我这辈子从来没用水清洁过口腔，倒是每天都有烧酒和葡萄酒从嘴里冲刷而过。'我确实没能从他的唾液中找到任何

东西，这跟其他人唾液的情况相同。我也同样将黏在他前面牙齿上的物质和他的唾液进行了混合，但除了少数一些迄今为止反复提到过的那种活体微小生物之外，我没有找到任何东西。但是，在那些从他的前牙缝隙（因为他的嘴里已经没有后牙了）中取到的物质里，我观察到了许多其他的微小生物，其中包括两个最小的种类。"

3）对自身牙菌斑样本进行的干预

"我故意三天没有清洁口腔，然后取了少量黏在我门牙上方和牙龈上的东西，我将它们与唾液以及干净的水混合在一起，发现里面有一些活的微生物。"

"此外，我把一些很浓的酒醋放进嘴里，咬紧牙关，让醋在牙齿之间来回流动了几次，接着用清水漱口三次。然后，我按照之前提到过的方法，再次从我的门牙和磨牙之间取了一些白色物质，与唾液以及干净的雨水混匀。我之前总是可以观察到数量惊人的活体微小生物，但它们主要存活在我从磨牙间取出的物质里，而且呈现出图1.2A所示形状的微小生物很少。我还把一点酒醋混合到唾液和水中，里面的小生物立刻死掉了。但是我猜想，我嘴里的醋并没有彻底穿透那些牢牢黏附在我的门牙和磨牙之间的物质，只是杀死了那些在比较靠外的白色物质里生活的微生物。"

4）关于口腔微生物的数量

"我的家里有几位女士，她们急切地想看看那些醋里的'小精灵'。有些人觉得她们看到的东西过于恶心，以至于让她们发誓以后再也不想食用醋。但如果未来有人告诉她们，在一个人口腔中，那些附着在牙齿上的脏东西里，生活着的小生物的数量比整个国家的人还多，那会发生什么？尤其那些从来不清洁口腔的人，他们嘴里散发出的气味是如此恶臭，几乎使人丧失与他们交谈的勇气。就我个人而言，从自身的情况判断，尽管我一直在按照之前描述过的方式清洁口腔，每天在我口腔里生活的微生物也依然应该比在荷兰联合王国里生活的人还要多。因为我在自己一颗被牙龈附盖着的后牙上看到了如马毛般粗细的白色物质，它们的样子看起来应该已经有些日子没被粗盐刷洗过了，里面一定存在着数量极其惊人的微小生物。我猜测，从这些白色物质里取出体积不超过沙粒的1%大小的部分，就应该可以识别出上千个活着的微生物。"

第 1 章 超有机体——微生物与宿主共生

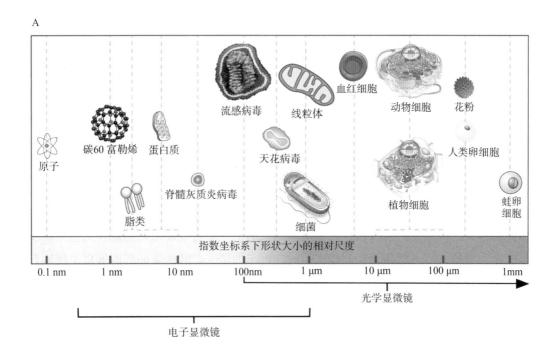

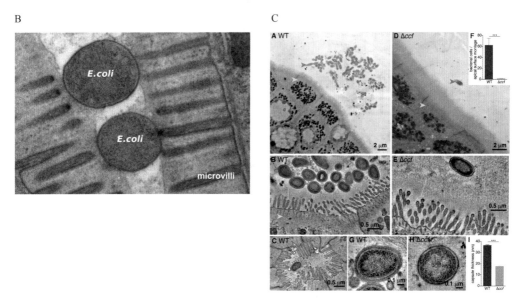

图 1.2　大部分细菌的体积比它们的宿主细胞小一个数量级

A.不同生物分子和细胞的尺寸图。B.在秀丽隐杆线虫的肠道中，大肠埃希菌细胞就像奶茶吸管中的木薯珍珠，它们的横截面直径约为 400 nm。C.野生型及 Δccf 突变型脆弱拟杆菌在单菌定殖小鼠的结肠中。注意，Δccf 突变型的细胞壁较薄。

图片来源：A. https://courses.lumenlearning.com/microbiology/chapter/types-of-microorganisms/.

B. 摄影者为 Shigeki Watanabe 和 Erik Jorgensen。

C. http://science.sciencemag.org/content/360/6390/795.long.

5

宏基因组学近20年来的发展代表了微生物学研究技术自显微镜发明以来的又一次重大飞跃，这使得我们能够更好地理解微生物世界（图1.3）。我们好奇又兴奋地将这一技术应用到了各式各样的样本中。无论多么复杂的样本，来自宿主来源的DNA（微生物的宿主通常是植物或动物），或是来自环境，我们都能对其全部DNA进行测序，然后通过组装测序得到的DNA片段来获得单独微生物的基因组信息，并对不同微生物在样本中的丰度进行定量。在高通量测序技术出现之前，基于传统微生物学以及之后出现的分子生物学技术，广大研究者们已经取得了很多优秀的成果。而依托高通量测序技术开展的宏基因组学作为一种新的研究方法，可以使得研究人员与医生不必再猜测培养条件并在培养皿中培养出每种微生物，就可以知道某处所有的微生物是什么，以及它们可以做什么。不同通量的测序平台以及生物信息学的进一步发展完善，使得有着不同预算规划或实验周期诉求的研究者和医生们都能够更便捷地开展宏基因组学研究。

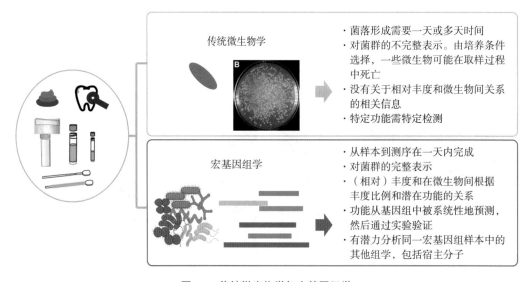

图1.3　传统微生物学与宏基因组学

如果每一步骤都进行定量，则可以获得绝对丰度（详见第3章）。在生态学研究中，相对丰度更为重要（详见第2章）。

我们对人体微生物的探索，将会带来更多的技术进步。而更好地关照生活在我们体内以及附着在我们体表的微生物，将能够帮助我们更好地应对全球人类群体层面的各种疾病。

示例 1.2　宏基因组，微生物群落，微生物组，傻傻分不清楚

宏基因组（Metagenome）：

Jo Hendelsman 博士在 1998 年关于土壤微生物的一篇论文中，引入了"宏基因组"的概念，意指一个环境样本中的全体基因分析研究。由于只有一小部分微生物能够被分离培养，她在文章中将全部土壤样本中的 DNA 都克隆进了大肠埃希菌的 BAC 文库（细菌人工染色体），然后做进一步的基因功能分析，而无须培养任何一种微生物。现在，随着高通量测序及生物信息学分析的成熟应用，我们使用"宏基因组学"一词来指代对任意样本中的微生物群落进行的无差别直接测序。

微生物群落与微生物组（Microbiota, Microbiome）：

关于"微生物组"，一个颇为流行的说法是这个名词是由诺贝尔奖得主 Joshua Linderburg 在 2001 年提出来的。从字面意义来看，"微生物群落"或"微生物丛"更多地关注由微生物组成的生态系统。而"微生物组"并不是一个基因组学时代（-ome）的产物，它的含义是包含在这个生态系统中的一切，毕竟该词的词尾是"生物群系"（biome）[6]。事实上，用来描述微生物社群关系的说法可以追溯到更早时候[6,7]。

早在 19 世纪末，维诺格拉茨基（Sergei Nikolaie-vich Winogradsky）就发现，当局部氧气被好氧硝化细菌（将铵盐氧化成亚硝酸盐，亚硝酸盐再氧化成硝酸盐）耗尽后，厌氧固氮型的巴氏梭菌就可以生活在由它的邻居创造的生态位中协助完成这两步硝化过程。由此，他在 1923 年提出应该研究自然环境状态下相互作用的微生物。

在他 1949 年关于纯培养的一篇论文中，有一些非常有趣的评论：

"……人工环境中的纯培养条件永远无法与自然环境中的培养条件相提并论……没有人可以否认这个观点：

一种微生物，如果被培育庇护在一个不仅没有任何竞争者存在，还供给了奢侈营养物质的环境中时，就成为了一种温室培养，它在短时间内就会被诱导成为一个新的种族，不经过专门的针对性研究就无法分辨出它原来是什么物种"[6,8]。

20 世纪 60 年代，使用特定的无菌无病或原动物模型成为了一种常规的实验室实践方案。鉴定筛选出来的微生物群落是否有助于维持动物模型的持续健康已经成为了一个常规指标：

"简而言之，赋予动物模型经过筛选的有助于维持它们持续健康的微生物群落，即赋予它们抵御在无菌环境之外一定会遇到的其他微生物攻击的能力"，Lane-Petter 在 1962 年写道[9]。1986 年，Linda R. Hegstrand 和 Roberta Jean Hine 发现无菌动物和常规饲养动物的下丘脑组胺水平存在差异，这是脑肠轴存在的早期例证[7]。

> 1988年，John M. Whipps在研究植物病害时写道：
> "研究生物控制系统时有一个便捷的生态学框架，这就是微生物组。微生物组可以被定义为占据了合理且明确的栖息地的特定微生物社群关系，而且这个栖息地具有独特的物化特性。因此，这个术语不仅指代涉及的微生物，同时也包括它们的活动区域。"[6,10]

1.2 人体内有多少微生物细胞？

细菌的大小通常位于较小的微米尺度上（图1.1），如果成丝状结构则另当别论。真菌通常比细菌大，而病毒通常比细菌小。一个自由生长的细菌，它的细胞的最小体积取决于细胞内部必需DNA及蛋白质的大小，而最大体积则是由维持该体积所需的核糖体数量决定的，但是核糖体也不能过多，否则会超出该细胞可容纳的体积（图1.4）[11]。有研究发现，在缺少营养物质时，大肠埃希菌的细胞质体积会立即减少17%。

而对于古菌，该原则同样适用。目前已知的最小古菌，其体积和最小的细菌相当，大约为 3.41×10^{-20} m³，基因组大小为 0.5 Mb，平均每个细胞约有 92 个核糖体[11]。

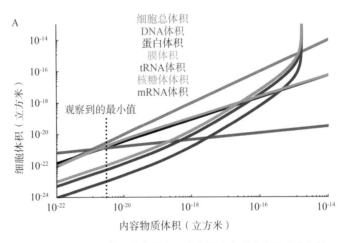

图1.4 决定细菌和古菌体积不会变得太大或太小的理论阈值

例如，一个典型的大肠埃希菌长度约2 μm，直径约0.5 μm，A.细菌各主要细胞器的体积依赖规律呈幂率（scaling）增长。B.细胞总体积与该细胞需要的全部细胞器的总体积之间的比较。C.根据细胞体积计算得出的细胞密度的幂率增长曲线，黑色曲线为计算出的细胞密度，红色曲线是作为参考的水的密度。

图片来源：Kempes CP, Wang L, Amend JP, et al., 不同细菌细胞组分的进化权衡. ISME J, 2016, 10:2145-2157. https://doi.org/10.1038/ismej.2016.21.

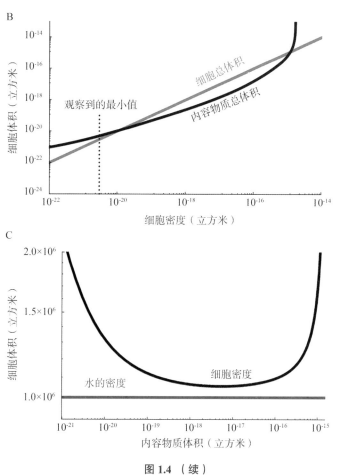

图 1.4（续）

提到大型动物，我们通常会想到老虎、大象或鲸鱼，但人类实际也属于地球上大型动物的一员。大型动物体重的对数，与寄生在动物们身上的微生物数量的对数之间，存在着一个有趣的线性关系（图 1.5A）。平均每克动物体重，对应存在着大约 3.4 亿个原核生物（细菌或古菌），而在动物们的体重中，有 0.34% 属于原核生物[13]。地球上微生物的总细胞数为 9.2×10^{29} ~ 31.7×10^{29}，其中合计数量为 2.1×10^{25} ~ 2.3×10^{25} 个细胞生活在动物身上。据估计，肠道（动物学意义上的肠道，即整个消化道）的体积随动物体重的增长的幂率系数是 1 ~ 1.08，而肠道的表面积随体重增长的幂律系数是 0.75。然而，每单位肠道体积或每单位肠道内容物质量的微生物数量变化可以有几个数量级，并且不同的动物，密切参与大量微生物活动的肠道区段也不同。

一个健康的成年人大约携带 3.8×10^{13} 个微生物细胞，其中绝大多数位于结肠（图 1.5B 和图 1.6）[14,15]。结肠中的微生物物种超过 600 种，每种细菌的数量在 1 ~ 10^{13} 个之间，因此对应的相对丰度在 10^{-13} ~ 0.3 之间不等（所有物种的相对丰度之和为 1）。皮肤拥有巨大的表面积，但其微生物群落在人体中相对简单（图 1.7A

和图 1.7B）[16,17]。

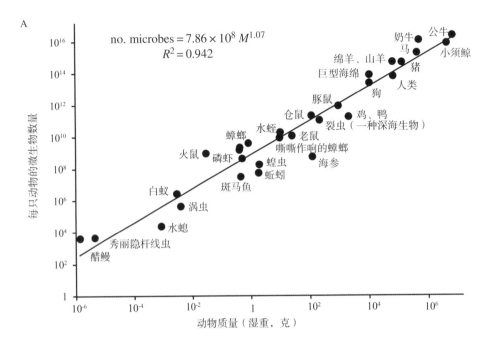

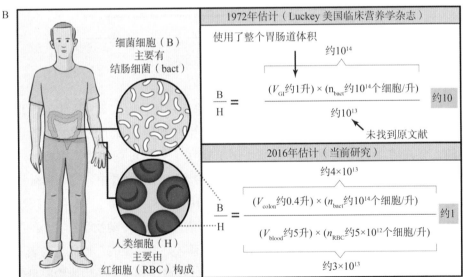

图 1.5 当前对成年人类体内的微生物数量的预估，以及不同体型动物间的总体趋势

A.每个动物包含的细胞数和其体重（含水，单位克）之间的关系（坐标系对数化）。B.由于结肠中微生物的密度要高得多，因此细菌细胞总数是根据成人结肠的一般体积而不是整个胃肠道的体积估算的。图片来源：A. Kieft TL, Simmons KA. 动物-微生物相互作用的异速生长和动物相关微生物的全球普查. Proc R Soc B Biol Sci. 2015. 282:20150702. https://doi.org/10.1098/rspb.2015.0702. B. Sender R, Fuchs S, Milo R. 我们真的人数众多吗？重新审视人类细菌与宿主细胞的比例. cell, 2016, 164：337-340. https://doi.org/10.1016/j.cell.2016.01.013.

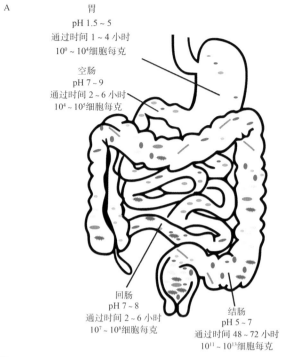

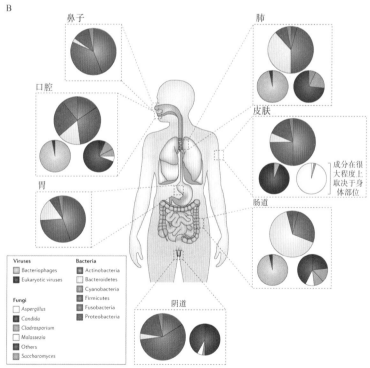

图 1.6 黏膜组织是人体共生微生物的主要栖息地

A. 处于消化道不同部位的微生物细胞估计数量，显著影响群落生存的因素，如pH及通过时间。（https://doi.org/10.1038/ismej.2012.6，）B. 不同部位的细菌（门）、真菌（属）和病毒。该图展示了处在不同外界环境下的身体不同部位（鼻腔、口腔、胃、肠道、阴道和肺）的细菌、真菌和病毒的相对丰度。细菌组成以6个最常见的门为代表，即放线菌（*Actinobacteria*）、拟杆菌（*Bacteroidetes*）、蓝细菌（*Cyanobacteria*）、厚壁菌（*Firmicutes*）、梭杆菌（*Fusobacteria*）和变形杆菌（*Proteobacteria*）。真菌组成包括曲霉（*Aspergillus*）、念珠菌（*Candida*）、枝孢霉（*Cladosporium*）、马色拉菌（*Malassezia*）和酵母（*Saccharomyces*）作为典型代表，其他类型的真菌被归纳为"其他"。病毒组成简单地分类为噬菌体或真核病毒。（https://doi.org/10.1038/nri3769）

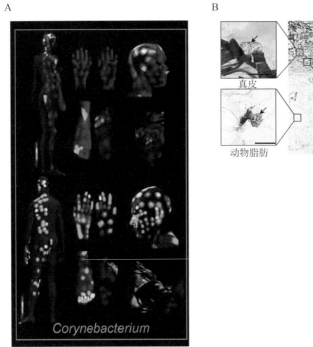
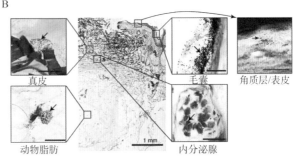

图 1.7　位于皮肤表面和内部的细菌

A. 棒状杆菌属（*Corynebacterium* spp.）在人类皮肤上的分布。B. 革兰氏染色法检测到的位于皮肤内的细菌。样本切除自黑色素瘤标本的正常区域。

　　肺部在拓扑学意义上是一个和外部连通的空间，有着更大的表面积和略显稀疏的微生物群落[18,19]。口腔和阴道中包含有大量细菌及少量病毒，有时候还会有真菌，但即使考虑上膀胱和子宫，它们中微生物的总和还是要比结肠中的微生物少 1～2 个数量级（图 1.8）。精液样本包含 10^6 ～ 10^7 个/mL 微生物[20]。在传统观念中认为不包含微生物的部位中（如肿瘤），也发现了微生物（图 1.9），不过数量不多。如果某一天，我们发现在人体的 10^{13} 个细胞中，每个细胞都包含两个细胞内微生物，我们可能就需要开始担心微生物总量的估计了（图 1.5）。

第 1 章 超有机体——微生物与宿主共生

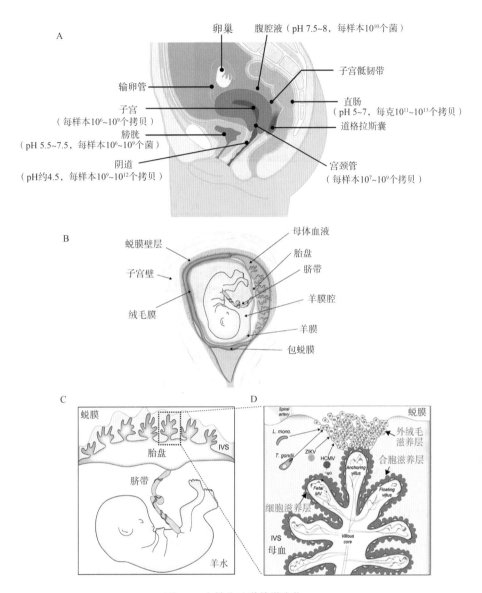

图1.8 女性生殖道的微生物

A. 对因非感染原因（例如子宫肌瘤）进行手术的未怀孕志愿者的生殖道不同区域的微生物数量的初步估计[21]。总的细菌基因组拷贝数是根据对样本中的乳酸杆菌进行qPCR计算出来的。上生殖道不再为酸性，这些乳酸杆菌在上生殖道中所占的比例跟在阴道、宫颈相比小很多。B~D. 对于胎盘不同位置微生物个数的估计，还需要更多研究（详见第3章），羊水样本通常仅适用于早产时。B. 妊娠期子宫腔内示意图。发育中的胎儿被包裹在羊膜腔内，被绒毛膜和羊膜包围，并被胎盘固定在母体蜕膜上（附着处的基底蜕膜处）。C. 当母体微血管系统建立之后，母体血液即通过螺旋动脉充盈胎盘表面的绒毛间隙（IVS）。D. 人血色绒毛状胎盘是由游离绒毛和锚定绒毛组成的绒毛树，通过绒毛外滋养层细胞的侵入，直接附着在基底蜕膜上的。人胎盘绒毛树被合胞体滋养细胞所覆盖，在这一层下面有一层细胞滋养细胞（在整个妊娠期间会变得不连续）。一些病原体，包括单核细胞增生李斯特菌（*Llisteria monocytogenes*）、刚地弓形虫（*Toxoplasma*）、人巨细胞病毒（*human lytomegaiovirus*，HCMV）和寨卡病毒（*zika virus*，ZIKV），被认为是在基底蜕膜复制后进入绒毛核的。

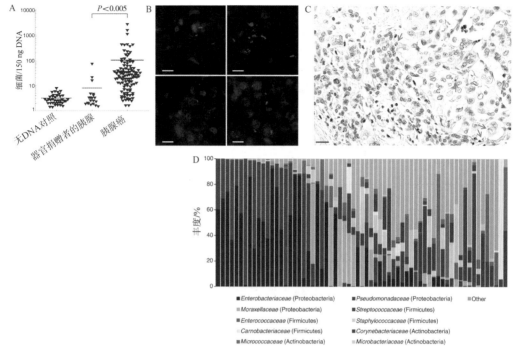

图1.9 通过多种检测方法，在胰腺导管腺癌（PDAC）中发现的细菌

A. 用细菌16s rDNA qPCR检测人胰腺肿瘤或器官捐赠者的健康胰腺组织中细菌存在情况。将细菌DNA插入人DNA中生成一条校准曲线，用于估计细菌数量。条形图代表平均值。B. 用荧光原位杂交检测人PDAC肿瘤（红色）中细菌16S rRNA基因序列。细胞核用4′,6-二脒基-2-苯基吲哚（DAPI）（蓝色）染色。展示了来自一个肿瘤的四张切片。比例尺：10 mm。C. 使用抗菌脂多糖抗体对人类PDAC肿瘤进行免疫组化分析。箭头指向肿瘤组织内的LPS染色。比例尺：20 mm。D. 65例人类PDAC肿瘤的系统发育类型分布。绘制了每个肿瘤的相对丰度（%）。

思考题1.1

（1）你认为哪些生理条件的改变，可能影响特定身体部位的微生物数量？

（2）在某个特定的身体部位，每天会损失多少个微生物细胞？

（3）每天损失的微生物细胞数目，是否相当于每天需要新分裂出来的微生物细胞数目？它们是否能被其他部位的微生物所补充？（更多内容见第2章和第4章）

思考题1.2

（1）你认为人体内的微生物种类会有几千种吗？

（2）你认为目前出现过的哪些数字可能被低估了？

1.3 人体中的病毒微粒

在很多类型的样本中，都已经检出了病毒。比如每克粪便中含有10^9个病毒样颗粒（virus-like particles，VLPs），每毫升尿液中含有 10^7 个病毒样颗粒，而每毫升唾液中含有 10^8 个病毒样颗粒[22]。

人体内常见的病毒为噬菌体，也有针对真核生物的病毒。指环病毒科（Anelloviridae）是一个无包膜的单链 DNA 病毒家族，拥有 2~4 kb 的环形基因组，通常在所有的主要黏膜部位以及血液和精液中都可以检测到。这个家族的病毒主要包括细环病毒（torque teno）、小细环（torque teno mini）病毒和中细环（torque teno midi）病毒。

1.4 其他物种中的微生物组

本书主要聚焦人类微生物与疾病的关系。在介绍探讨了基本原理后，读者应该可以在应用宏基因组学研究探索动物甚至植物领域相关具体问题的过程中汲取灵感（图 1.10）[23]。

在特定的宿主谱系中，微生物组的主要组分可能会出现极端变化。蝉这种昆虫通常是依赖必需的内共生菌 *Sulcia* 和 *Hodgkinia* 存活下来的，但在许多日本蝉中，*Hodgkinia* 已经被一种真菌所取代，这种真菌很可能是从寄生在蝉身上的 *Ophiocordyceps* 真菌招募转化来的。这种内共生真菌可以编码所有 B 族维生素和氮循环的通路，合成所有必需和非必需的氨基酸，比 *Hodgkinia* 提供的组氨酸和蛋氨酸合成更加灵活完善[25]。而果蝇（*Drosophila melanogaster*）会积极地寻找食物来补充它们的肠道细菌，并且能通过早期接触细菌改变自身共生微生物的偏好[26]。

在传统微生物学中，pH、氧气、温度、矿物质、碳源和氮源等所有因素都需要被摸索确认清楚，以便对生长产生正向影响（图 1.11）[27]，而有些细菌自身是可以固定二氧化碳或者氮的。在一个微生物群落中，无论处于宿主的表面还是内部，都既是机会也是挑战（详见第 2 章）。微生物会向宿主提供一系列服务，如竞争性排斥病原体、消化复杂的有机基质、提供营养物质和生长因子、刺激发育、影响行为等。这些服务都具备演化保守性，可以操纵宿主为微生物提供更好的生存条件[28-31]。

在营养贫乏（如深海沉积物）环境中，细菌需要一千多年才能复制一次。相比而言，尽管不同类群之间可能存在差异，人类共生微生物的复制速度已经很快了。

1.5 来自远古的微生物组

古基因组学告诉了我们许多关于现代人类和其他物种的进化和迁徙的信息。而微生物 DNA 片段也可以从牙齿化石和粪化石（化石粪便）中被提取（图 1.12、图 1.13），这些样本提供了令人兴奋的数据，可以让我们一窥微生物组的进化演变。

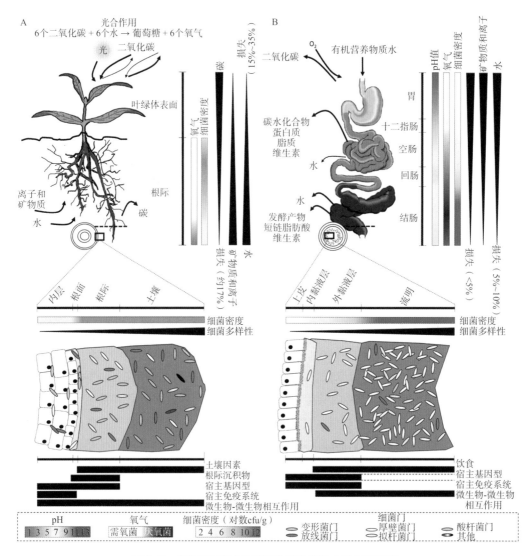

图 1.10 植物根系和人体肠道在养分吸收中的生理功能，微生物群落组成的空间方面，以及驱动群落形成的因素对比

植物根部微生物区系（A）和人体肠道微生物区系（B）的空间上的分区情况。图中显示了主要的营养物质通量，以及与细菌密度有关的pH和氧气的梯度。下图显示了微生物群沿着肠道内腔上皮或根部的土壤-内皮呈连续状的分区情况。对于每个分区，对应到肠道和植物根部，标示了细菌的密度、多样性和细菌的主要代表门。在这些不同的分区中推动社群建立的主要因素用黑条表示。这幅内脏图来自Tsabouri等（2014），得到了出版商的许可。

列文虎克的发现使我们对发达国家卫生习惯发生了怎样的变化有了更多了解（示例1.1）。而这些远古时期的牙齿、粪便和环境样本则可以向我们揭示当时的饮食习惯和微生物。来自西班牙一个洞穴的尼安德特人粪化石样本显示，当时的人肠道微生物组中有许多与人类现在的样本相同的属[34]。可能包括病毒、细菌、真菌和寄生虫在内的病原体，都在影响塑造人类基因组和微生物组的过程中发挥了重要作用。这些历史故事还有待读者自己去进一步探索。

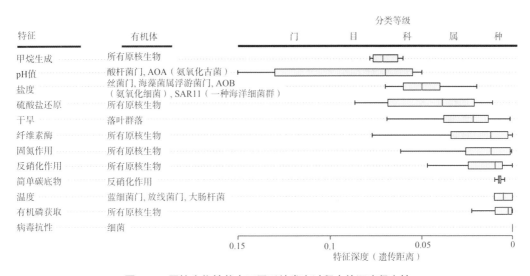

图1.11 原核生物性状在不同系统发育过程中的深度保守性

分支深度的箱线图,其中每个分类群始终共享一个特征,这个特征被测量为一个分支距根节点的遗传距离(横轴;通常是16S rRNA基因)。对于某些性状,分布基于几项研究,每项研究都有一个估计值。对于其他特征,原始数据的作者报告了单项研究计算出的分布。作为比较,作者在纵轴上显示了粗略的分类级别。

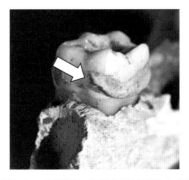

图1.12 下颌磨牙上一个凹环内的龈上牙结石标本(英国,约克郡)

图1.13 来自白垩纪的粪化石(化石粪便)中类似微生物的形态,不过DNA无法保存这么久

A. 美国蒙大拿州东部上白垩纪地狱溪组河流沉积物中保存的三角龙(sterrhophus Marsh)遗址的粪化石。宿主动物未知。粪化石中含有少量微小的骨骼或牙齿碎片、干酪化的植物残体(花粉、孢子、孢子囊、角质层等)、可能是来源于真菌的菌丝,以及高孔隙度且纹理细密的基质里的微小细碎矿物颗粒。B. 分叉的真菌或细菌菌丝,以及许多大小和形状上类似细菌小球形物体。比例尺10 μm,M=硅酸盐矿物颗粒。C. 粪化石基质的结构显示有空心薄壁矿物球体,其中一些具有双层壳,两壳之间有薄空隙,可能是以前细菌细胞壁的位置。比例尺2 μm,M=硅酸盐矿物颗粒。

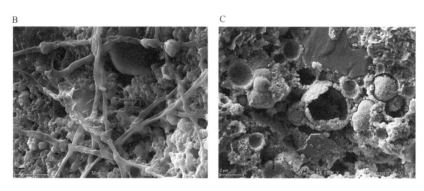

图1.13 （续）

1.6 总结

宏基因组学正处于人类微生物组研究的黄金时期。本章从列文虎克对牙菌斑中微生物的观察（示例1.1）开始，介绍了在微生物学技术发展背景下，对人类微生物组的宏基因组学研究工作。微生物细胞的数量随动物生物量的变化而变化。目前对一个成年人体内 3.8×10^{13} 个微生物细胞的估计是基于结肠的体积，而结肠正是绝大多数人体共生微生物定植的地方。其他部位，如上皮细胞、组织或体液中的微生物数量则较少。正如在示例1.2中详述的那样，我们将在本书中使用微生物组这一概念来代表微生物群落或微生物丛。

除了技术和成本之外，我们对微生物组的研究是不受限制的。本书并不是按照传统的人类微生物组所在的身体部位来划分章节的，我们希望能够为未来数年里的新发现提供空间。微生物细胞的数量，或者每个部位的微生物生物量将是宏基因组学研究的一个重要考虑因素（第3章、第4章、第5章）。在第2章介绍了生态学原理之后，本书继续介绍了设计宏基因组研究的更多实用知识（第3章、第4章），不同的分类精度（第5章），以及我们如何得出微生物组在疾病中发挥作用的因果结论（第6章）。然后，我们就可以把这些知识应用到临床实践（第7章）和更长期的健康管理（第8章）中。

原著参考文献

[1] Lane N. The unseen World: reflections on Leeuwenhoek (1677) 'Concerning little animals.'. Philos Trans R Soc B Biol. Sci 2015;370:20140344. https://doi.org/10.1098/rstb.2014.0344.

[2] Cocquyt T, Zhou Z, Plomp J, van Eijck L. Neutron tomography of Van Leeuwenhoek's microscopes. Sci Adv 2021;7. https://doi.org/10.1126/sciadv.abf2402, eabf2402.

[3] van Leeuwenhoek A. The collected letters of Antoni van Leeuwenhoek. Amsterdam: Swets and Zeitlinger; 1952.

［4］Woese CR, Kandler O, Wheelis ML. Towards a natural system of organisms: proposal for the domains Archaea, Bacteria, and Eucarya. Proc Natl Acad Sci U S A 1990;87:4576–4579. https://doi.org/10.1073/pnas.87.12.4576.

［5］Handelsman J, Rondon MR, Brady SF, Clardy J, Goodman RM. Molecular biological access to the chemistry of unknown soil microbes: a new frontier for natural products. Chem Biol 1998;5:245–249.

［6］Goin J. Microbiomes: an origin story. ASM.org, https://asm.org/Articles/2019/March/Microbiomes-An-Origin-Story#; 2019.

［7］Prescott SL. History of medicine: origin of the term microbiome and why it matters. Hum Microbiome J 2017;4:24–25. https://doi.org/10.1016/j.humic.2017.05.004.

［8］Dworkin M. Sergei Winogradsky: a founder of modern microbiology and the first microbial ecologist. FEMS Microbiol Rev 2012;36(2):364–379. https://doi.org/10.1111/j.1574-6976.2011.00299.x.

［9］Lane-Petter W. The provision and use of pathogen-free laboratory animals. Proc R Soc Med 1962;55:253–263. https://doi.org/10.1177/003591576205500402.

［10］Whipp, et al. Fungi in biological control systems. Manchester University Press; 1988.

［11］Kempes CP, Wang L, Amend JP, Doyle J, Hoehler T. Evolutionary tradeoffs in cellular composition across diverse bacteria. ISME J 2016;10:2145–2157. https://doi.org/10.1038/ismej.2016.21.

［12］Shi H, Westfall CS, Kao J, Odermatt PD, Anderson SE, Cesar S, et al. Starvation induces shrinkage of the bacterial cytoplasm. Proc Natl Acad Sci U S A 2021;118. https://doi.org/10.1073/pnas.2104686118.

［13］Kieft TL, Simmons KA. Allometry of animal–microbe interactions and global census of animal-associated microbes. Proc R Soc B Biol Sci 2015;282:20150702. https://doi.org/10.1098/rspb.2015.0702.

［14］Sender R, Fuchs S, Milo R. Are we really vastly outnumbered? revisiting the ratio of bacterial to host cells in humans. Cell 2016;164:337–340. https://doi.org/10.1016/j.cell.2016.01.013.

［15］Sender R, Fuchs S, Milo R. Revised estimates for the number of human and bacteria cells in the body. PLoS Biol 2016;14. https://doi.org/10.1371/journal.pbio.1002533, e1002533.

［16］Bouslimani A, Porto C, Rath CM, Wang M, Guo Y, Gonzalez A, et al. Molecular cartography of the human skin surface in 3D. Proc Natl Acad Sci U S A 2015;112:2120–2129. https://doi.org/10.1073/pnas.1424409112.

［17］Byrd AL, Belkaid Y, Segre JA. The human skin microbiome. Nat Rev Microbiol 2018;16:143–155. https://doi.org/10.1038/nrmicro.2017.157.

［18］Hasleton PS. The internal surface area of the adult human lung. J Anat 1972;112:391–400.

［19］Man WH, de Steenhuijsen Piters WAA, Bogaert D. The microbiota of the respiratory tract: gatekeeper to respiratory health. Nat Rev Microbiol 2017;15:259–270. https://doi.org/10.1038/nrmicro.2017.14.

［20］Hou D, Zhou X, Zhong X, Settles ML, Herring J, Wang L, et al. Microbiota of the seminal fluid from healthy and infertile men. Fertil Steril 2013;100:1261–1269. https://doi.org/10.1016/j.fertnstert.2013.07.1991.

［21］Chen C, Song X, Wei W, Zhong H, Dai J, Lan Z, et al. The microbiota continuum along the female reproductive tract and its relation to uterine-related diseases. Nat Commun 2017;8:875. https://doi.org/10.1038/s41467-017-00901-0.

[22] Liang G, Bushman FD. The human virome: assembly, composition and host interactions. Nat Rev Microbiol 2021. https://doi.org/10.1038/s41579-021-00536-5.

[23] Hacquard S, Garrido-Oter R, González A, Spaepen S, Ackermann G, Lebeis S, et al. Microbiota and host nutrition across plant and animal kingdoms. Cell Host Microbe 2015;17:603–616. https://doi.org/10.1016/j.chom.2015.04.009.

[24] Tsabouri S, Priftis KN, Chaliasos N, Siamopoulou A. Modulation of gut microbiota downregulates the development of food allergy in infancy. Allergol Immunopathol (Madr) 2014;42(1):69–77. https://doi.org/10.1016/j.aller.2013.03.010.

[25] Matsuura Y, Moriyama M, Łukasik P, Vanderpool D, Tanahashi M, Meng XY, et al. Recurrent symbiont recruitment from fungal parasites in cicadas. Proc Natl Acad Sci U S A 2018;115:5970–5979. https://doi.org/10.1073/pnas.1803245115.

[26] Wong ACN, Wang QP, Morimoto J, Senior AM, Lihoreau M, Neely GG, et al. Gut microbiota modifies olfactory-guided microbial preferences and foraging decisions in drosophila. Curr Biol 2017;27:2397–2404.e4. https://doi.org/10.1016/j.cub.2017.07.022.

[27] Reimer LC, Vetcininova A, Carbasse JS, Söhngen C, Gleim D, Ebeling C, et al. BacDive in 2019: bacterial phenotypic data for High-throughput biodiversity analysis. Nucleic Acids Res 2019;47:631–636. https://doi.org/10.1093/nar/gky879.

[28] Fraune S, Bosch TCG. Why bacteria matter in animal development and evolution. Bioessays 2010;32:571–580. https://doi.org/10.1002/bies.200900192.

[29] McFall-Ngai M, Hadfield MG, Bosch TCG, Carey HV, Domazet-Lošo T, Douglas AE, et al. Animals in a bacterial world, a new imperative for the life sciences. Proc Natl Acad Sci U S A 2013;110:3229–3236. https://doi.org/10.1073/pnas.1218525110.

[30] King KC, Brockhurst MA, Vasieva O, Paterson S, Betts A, Ford SA, et al. Rapid evolution of microbe-mediated protection against pathogens in a worm host. ISME J 2016;10:1915–1924. https://doi.org/10.1038/ismej.2015.259.

[31] Sherwin E, Bordenstein SR, Quinn JL, Dinan TG, Cryan JF. Microbiota and the social brain. Science 2019;366. https://doi.org/10.1126/science.aar2016, eaar2016.

[32] Martiny JBH, Jones SE, Lennon JT, Martiny AC. Microbiomes in light of traits: a phylogenetic perspective. Science 2015;350:9323. https://doi.org/10.1126/science.aac9323.

[33] Jørgensen BB, Marshall IPG. Slow microbial life in the seabed. Ann Rev Mar Sci 2016;8:311–332. https://doi.org/10.1146/annurev-marine-010814-015535.

[34] Rampelli S, Turroni S, Mallol C, Hernandez C, Galván B, Sistiaga A, et al. Components of a Neanderthal gut microbiome recovered from fecal sediments from El salt. Commun Biol 2021;4:169. https://doi.org/10.1038/s42003-021-01689-y.

[35] Enard D, Petrov DA. Evidence that RNA viruses drove adaptive introgression between Neanderthals and modern humans. Cell 2018;175:360–371. https://doi.org/10.1016/j.cell.2018.08.034.

[36] Rasmussen S, Allentoft ME, Nielsen K, Orlando L, Sikora M, Sjögren KG, et al. Early divergent strains of Yersinia pestis in Eurasia 5,000 years ago. Cell 2015;163:571–582. https://doi.org/10.1016/j.cell.2015.10.009.

第 2 章

微生物群落生态

摘　要：本章将从微生态学的角度，探讨微生物群落的多样性、营养级等概念。在某些疾病进展过程中，患者的微生物群落可能呈现更高的多样性，这种情况更常见于肠道微生物之外的部位。例如大部分绝经后的女性，由于阴道酸性下降，原本由单一菌种主导的微生物群落，会变得更加复杂多元。本章还将简要介绍营养和选择压力对于皮肤和口腔微生物的影响。黏液素、免疫球蛋白和血液抗原聚糖等因素，都能影响微生物的觅食和定位。升结肠部位是食物残渣进入大肠后主要的生物反应器，而微生物在后续的结肠部位可以继续发酵，直到底物被完全消耗或者微生物已被排出体外。人的基因差异、个人生活经历以及心跳节律，都会对微生物群落的构成产生影响，而研究更多的是营养及免疫反应和菌群的关系。人类微生物中的营养级结构，需要考虑个体在不同场景下的身体状况。在不同状况下，微生物面对着不同的可用分子和进化优势。

关键词：微生物群落，微生物多样性，食物等级，塞伦盖蒂（Serengeti）法则，皮肤微生物，口腔微生物，肠道微生物

2.1　宏观生态学中的营养级

在一个经典的宏观生态群落中，太阳是唯一的能量来源，只有能够进行光合作用的微生物和植物才能利用这些能量，并通过各个营养级逐级积累生物质（图 2.1）。在食物链中，每个营养级的动物都是捕食者，它们在生态金字塔中占据不同的位置。维持生态系统的整体多样性，并不意味能代替对生态系统中进行特定功能的管理[1]。

近 20 年来，随着对共生微生物研究的复兴，微生态、多样性等术语也越来越受到关注。然而，我们对人体微生物组成的生态系统仍然知之甚少。人体内的微生物除了依赖宿主的分子和未消化的食物为生外，还有一些可以是部分或完全的化能自养菌（译者注：一类直接利用无机物来产生自身生长所需的能量和有机物的微生物），它们通过 CO_2 固定或厌氧硝酸盐呼吸来获取能量。微生物和宿主之间的分子交换，大部分无法直接观测，需要等待新的技术突破来揭示其奥秘。

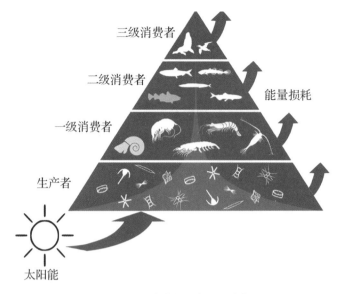

图 2.1 宏观生态学中,食物链的范例

来自太阳的能量,经由生产者(植物或微生物)被转换为生物量,每一层消费者都基于上一层提供的能量。

2.2 微生物群落稳定性、多样性及丰富度

类比宏观生态系统,一个流行的观点是,微生物群落更多样化,则更有可能保持稳定,甚至被视为是更健康的。样本内部的 α 多样性,通常根据香农指数计算,考虑物种的丰富度(richness)和相对丰度的均匀程度,因此能反映出分类单元的数量和相对多样性。然而,低丰度的微生物,可能稳定地生活在某个生态位中(不直接与更高丰度的微生物竞争),或有一个稳定的来源,这使得我们不必担心其在某处有灭绝风险。

对于细菌,大部分的功能特征是由属水平决定的(见第一章图 1.11,对于分类关系见第 5 章)[2]。有研究表明,不论起始状态如何,在同一食物孵育后,土壤或叶片的微生物种群会形成在科分类单元下相同的分布;而同一科内部的菌种可以大不相同,稀有的菌种还可以共存[3]。尽管合作可以提高效率,但理论分析表明,竞争才能增加微生物群落的稳定性并有利于宿主[4]。非特异性的交叉供应网络,而不是特定的细菌对之间的互动,能够使竞争中的微生物生态系统稳定[3]。空间结构和微生物种间的弱联系,也有助于提高生态系统的稳定性[4]。在小鼠肠道中,由于空间的异质性,噬菌体和作为其猎物的目标菌能够共生[5]。

大多数情况下,来自发达国家的人类粪便微生物的多样性更高,这意味着有更多细菌来自厚壁菌门(Firmicutes)而不是拟杆菌(Bacteroidetes)、放线菌(Actinobacteria)或变形杆菌(Proteobacteria)。属于厚壁菌门的多种细菌,负责人类肠道中的发酵

过程，其中涉及非特异性的营养共生及空间分异。在未进入农业社会的人群中，例如当代亚马孙森林中生活的居民，以及哈扎（Hadza）的采集狩猎部落，螺旋体（Spirachaetes）、疣微菌门（Verrumicrobia）和其他分类单元的微生物（示例2.1、图2.2），是如何导致肠道微生物多样性更高的机制依然未知。在这样彼此之间密切交往的社群中，人们的肠道微生物的β多样性较低，即个体之间的相似性较高，而不受年龄的影响。

示例2.1 密螺旋体，进入农业社会前的"肠道菌群主力"？

在发达国家，普雷沃菌（*Prevotella copri*）是一种已被认定为比拟杆菌（*Bacteroides* spp.）更为罕见的肠道细菌；而在发展中国家的部分人群中，前者的相对丰度却高达50%左右[110-112]。然而，在狩猎采集社会中，密螺旋体可能是肠道菌群的基石。在大猩猩和狒狒等非人灵长类动物的肠道中，也发现了密螺旋体[113]。密螺旋体在人类和大猩猩的肠道菌群中都呈现出季节性的变化，其变化趋势与纤维素摄入量增加呈正相关[114,115]。密螺旋体能够降解木聚糖纤维素（xylan cellulose），这种能力曾在白蚁的后肠中被发现[116]。而拟杆菌门的细菌，主要负责人体肠道和牛瘤胃中木聚糖的水解。烹饪后再冷却使高直链淀粉（又收缩形成抗性淀粉）的消化速度减慢，抗性淀粉会增加密螺旋体的丰度，同时降低普雷沃菌属的水平[117,118]。

螺旋体门广泛存在于多种动物体内[113,119]。根据英国双胞胎队列的宏基因组数据，螺旋体门似乎是可遗传的。但考虑到在现代人的样本中，我们只针对这一门的菌检测了不足10个基因，因此"螺旋体门因人类宿主的基因具有遗传性"这一说法还缺乏充分的证据[120]。

肠道中的密螺旋体，并不同于著名的牙周致病菌齿垢密螺旋体，也不同于性传播的到梅毒螺旋体。产琥珀酸密螺旋体（*T. succinifaciens*）已在当今多个偏远村落，以及1000年前的墨西哥人样本中被检测到[96]。产琥珀酸密螺旋体不能发酵氨基酸，但能固定CO_2生成琥珀酸。乙酸是通过还原性的产乙酸过程，由氢气和二氧化碳生成[121]。进食是导致体内二氧化碳水平随昼夜节律变化的主要因素[122]。普雷沃菌和拟杆菌，除了能够产生琥珀酸酯之外，还能生成丙酸盐、丙酸酯被用于肠道糖异生[123,124]。

基因或物种的丰富度，只是计算了每个样本中的基因或物种种类的数目。丰富度的增加，可能是由样本中丰度较低（即个体数目较少）的微生物引起的，这些低丰度的菌对于衡量每个分类单元均匀程度的α多样性的贡献很低。同时，检测到低丰度的微生物，也依赖于足够的测序深度（见第3章）。

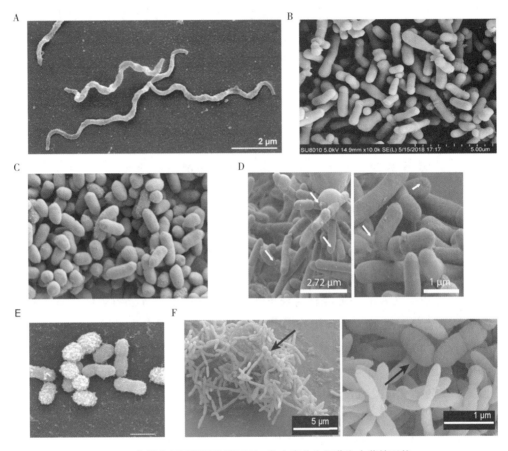

图2.2　扫描电子显微照片显示了一些人类共生细菌和古菌的形状

A.产琥珀酸密螺旋体；B.动物双歧杆菌乳亚种；C.多形拟杆菌；D.普氏栖粪杆菌（普拉梭菌）；E.阿克曼氏菌属；F.微小克里斯滕森菌与古细菌史密斯甲烷短杆菌共培养。箭头表示史密斯甲烷短杆菌

对于肠道（粪便）微生物，已知炎症性肠病与肥胖和肠道微生物的α多样性较低相关，而结直肠癌以及和便秘有关的疾病其肠道丰富度常常被发现较高，可能包含来自口腔和其他位置的菌群。对于牙齿和阴道，较低的多样性意味着更健康，而相对应的是牙周炎和细菌性阴道炎时多样性更高（表2.1）。人类生殖道的微生物多样性偏低，其中90%以上是单一种乳酸菌。在哺乳动物中，只有人类生殖道微生物由乳酸菌占主导地位，这在进化上是一个值得探索的现象。根据体外发酵研究，单一种群可能导致pH显著偏离中性，而在菌群多样性增加时这一趋势则会逆转。在生育年龄，人类生殖道的pH小于4.5（由于乳酸菌发酵糖原），而在绝经后，pH为7.0，细菌绝对数量是更少的（虽然多样性可能看上去高了）。

婴儿从母亲的生殖道和粪口途径获得微生物，之后通过母乳喂养和皮肤接触继续获得微生物（图2.3，详见第8章）。环境中来自其他家庭成员和宠物的微生物，包括产芽孢厌氧菌在内，也可以在婴儿体内定居。在母乳喂养结束后，上述因素很可能

表 2.1 疾病和微生物多样性

疾病	样本	相比健康对照菌群多样性变化	相比健康对照菌群丰富度变化	参考文献
超重	粪便	减少	在特定基因上减少+	[14-16]
克罗恩病	黏膜活检，粪便	减少	减少	[17-21]
溃疡性肠炎	粪便	在部分环境下减少	在部分环境下减少*	[20,22]
结直肠癌	黏膜活检，粪便	增加	增加	[23,24]
精神分裂	粪便	增加	增加	[25]
抑郁症	粪便	无差异	未报道	[26]
乳腺癌	粪便	在未绝经的患者中增加	在未绝经的患者中增加	[27]
间接母乳喂养（使用泵而非吸入进食母乳）	母乳	减少	减少	[28]
黑色素瘤	癌组织	未报道	增加	[29]
牙周炎	牙菌斑	增加	增加	[30]
细菌性阴道炎	生殖道拭子	增加	增加	[31]
早产	生殖道拭子	增加	增加	[32]
男性不育	精液	增加	未报道	[33]

该表并不全面，只列出各种可能性的范例。菌群多样性可以在物种或属的水平上衡量。+表示由人类肠道芯片（HITChip）检测的宏基因组而非核糖体16S得出。*表示可能由结肠镜检查导致的变化。多样性计算，按逆辛普森指数替代香农指数。

A 婴儿阶段
微生物来自母亲食物/水环境，包括人和宠物，在身体各部位微生物依次替换

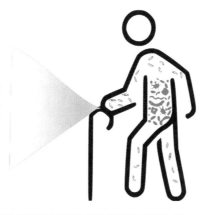

B 老年阶段
微生物来自食物/水环境，包括人和宠物，原有微生物在体内不同部位间迁徙与演替

图 2.3 人体共生微生物群落在各阶段的选择性来源

A. 婴儿阶段；B. 老年阶段

促进了婴儿肠道的微生物多样性增加[35]，断奶反应以及维生素 A 和短链脂肪酸共同标志着黏膜免疫系统的发育[36]。至少对于小鼠，在断奶前的关键时期引入固体食物，有助于降低过敏风险[36,37]。除了历史特异性，以及从母亲继承的微生物（图 2.4），婴儿阶段是全身免疫发育的一个关键阶段，免疫系统的发育将决定哪些新旧微生物能够生存。最近的"宏基因组 + 基因组"关联分析，都没有在样本中纳入婴儿及儿童，但理应考虑到自身基因的影响（示例 2.2）。

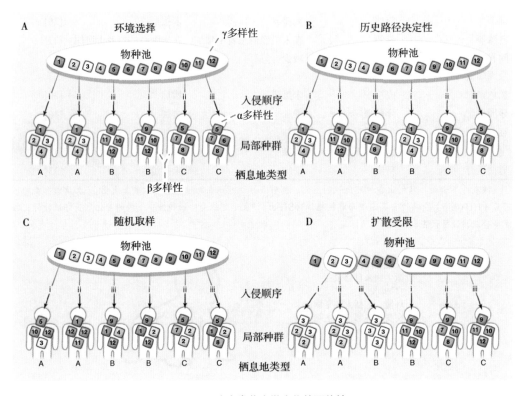

图 2.4　组建人类共生微生物的可能情况

不同的菌群集结组装场景可能导致人类微生物群中的丰度差异。图 A~C 中的每个小组，展示了从一个现有可行的物种池中组成局部群落的方式。每个局部种群可以接触到所有可用的寄生菌，但是接触的顺序不尽相同。在 A 中，当地物种组成主要由环境选择决定：无论接触顺序如何，初始条件相似的生境都有相似的菌群组合。在 B 中，相反的情况成立：不管最初的生境条件如何，历史的偶然性（如接触物种的时间和顺序上的差异）决定了物种组成。在 C 中，栖息地和历史都无关紧要：局部群落通过从物种池中随机抽取的物种组合而成。D 扩散障碍导致特定群落聚集在不同的物种池中。对于每个物种池，相应群落可以像 A、B 或 C 那样组合。A 表明了 3 种不同多样性指标的含义：γ 多样性指的是"区域"物种库（即通过扩散连接的局部群落的总多样性）；β 多样性指的是局部群落之间的差异（物种更替）；α 多样性指的是局部群落内部的多样性。尽管上述的多种场景可能适用于任何现实世界的环境，但其中一种可能会占主导地位。例如，寄生在身体不同部位的菌群差异可以用环境选择来最佳解释，同一栖息地的兄弟姐妹之间的差异最适合用历史上的偶然性来进行解释，断奶前的同卵双胞胎之间的差异突显了随机性的作用，剖宫产和阴道分娩的新生儿之间的差异可以用扩散过程受到限制来解释。

正常人在健康衰老过程中，肠道微生物群落的丰富度增加，原本主导的分类单元所占的百分比下降[38-40]。但我们不清楚这些数量变化，是由于免疫系统变弱（IgA抗体水平降低以及肠道蠕动减慢导致肠道微生物滞留，见第2章第6节），或是由于随着身体的衰老，微生物在身体的其他部分增多。生殖道、肠道和口腔的微生物都与激素存在相关性[12,41]。主导的菌数量减少，可能解除了丰度较低物种的生长限制。对于衰老的免疫系统，是否要消灭入侵的微生物，会是一个困难的抉择。部分粪便微生物中的疾病标志物，针对诸如结直肠癌、肝脏、心血管疾病的标志物，可能来自口腔，这些标志物在肠道中的丰度和人体的基因差异有关[42]。

> **示例2.2　安东尼·范·列文虎克在研究姜水时对黏液、黏膜个体差异的一些思考**
>
> 在1676年10月9日写给伦敦皇家学会的一封长信中，他在信中写下了他的一些有趣的思考："因此，也可能发生一个身体比另一个具有更大的内部热量或运动，或者有些肠道被柔软的薄黏液覆盖，有些则被厚厚的黏液覆盖，还有些被坚硬或硬质的黏液覆盖，因此，一些人可能通过非常温和的药物得到净化，而另一些人则需要强烈而猛烈的药物。但我同样愿意相信存在极小的、刺激性的粒子，它们能随着食物一起穿透黏液，直到它们进入肠道的球状体等才发挥作用。请原谅我再次谈论我并不了解的事物，但请记住我是谁，我是什么。"[8] 这封信在皇家学会的版本中并未全文发表。

2.3　皮肤微生物的几种不同生境

尽管在表皮下也发现了微生物的存在（第1章，图1.6），但我们对其知之甚少。故只讨论生活在我们身体外部的皮肤微生物。其主要生境类型包括：皮脂腺（油脂、蜡脂）、潮湿的、干燥的皮肤和足部皮肤组织，对应于非常不同的群落（图2.5）。不论对于男性和女性，随着年龄的增长，皮肤的酸性变弱。从青春期到中年，大量的皮脂分泌为痤疮角质细菌（又称痤疮丙酸杆菌）提供了独特的生态位。因此，每当在上生殖道或大脑中看到痤疮角质细菌的DNA时，我们就会好奇它在那里以什么脂质为食，或者它是否已经被噬菌体杀死了。表皮葡萄糖球菌和多种棒状杆菌在皮肤各部位均有分布。除了噬菌体，多瘤病毒，如潜在致癌的默克尔细胞多瘤病毒和乳头瘤病毒也经常被检测到[48]。真菌马拉色菌可以在毛囊以及在表皮被检出[48]。在前臂等干燥部位，则不能维持较高的微生态生物量，而且皮肤微生物组趋于多样化和可变化[49,50]。

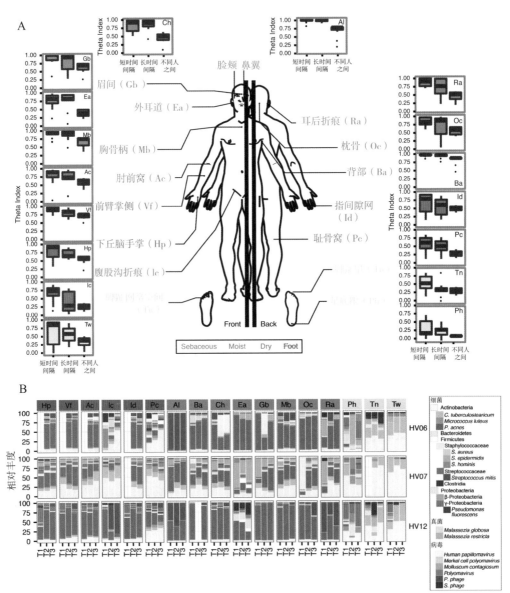

图 2.5　皮肤微生物的不同类型和它们的时序稳定性

A. 该研究中17种皮肤微生物的采样点，以及它们在人体中的位置。这些采样点代表了四种微环境：油脂（绿色）、干燥（蓝色）、湿润（红色）和足部（黑色）。B. 对于所有采样点，箱线图中的Yue-Clayton theta值，代表时间序列中，不同样本间的相似度，代表了皮肤上不同位点的特征。长的箱体代表采样间隔1年，短的代表间隔1个月。与之对比的，个体间的值代表不同个体间的平均距离。黑线代表中位数，箱体包含了最高25%及最低25%的范围。柱状图中的颜色代表采样点的特征。对于所有的个体之间和个体之内的对比，同一个体的P值，预期小于0.05。大部分常见皮肤细菌、真菌和病毒，在三个代表性个体中的相对丰度。T1、T2和T3代表时间序列的次序。来源：Cell, 2016, 165:854-866. https://doi.org/10.1016/j.cell.2016.04.008

粪便微生物不符合Hubbel的中性选择理论，但某些生殖道、皮肤及呼吸道微生物群落似乎符合这一理论。中性选择理论，相比传统的生态位理论，强调从一个源群落中的随机抽样（图2.4），之后伴随着在争夺局部生存空间中的随机增长和死亡的动态过程。

化学制品在我们身体中的停留时间比预期的更长，尤其在它们成为微生物的食物来源以前。在一项研究中，两位志愿者被要求 3 天内避免淋浴、使用卫生或美容用品后，结果发现大部分已知的代谢产物，都可以追溯到个人卫生或美容产品的原料（对于部分未知的代谢物，是否由微生物代谢产生，目前尚未研究）[53]。例如，表面活性剂 C12 十二烷基醚硫酸酯和可可酰胺丙基甜菜碱，防晒成分阿伏苯宗和十八烷基[53]。

思考题 2.1

（1）在出汗过程中，你认为皮肤微生物会发生什么变化？
（2）不同身体部位的皮肤微生物的构成和维持机制，有哪些不同之处？
（3）根据皮肤微生物的分类单元构成以及功能组成，能够推测出哪些个人习惯？

2.4 影响口腔微生物的因素

牙齿、口腔黏膜或口腔角质化上皮中，会有微生物直接或间接（通过其他微生物或宿主环境成分）地黏附在表面（图 2.6）。它们的生活，受到每天的口腔清洁、饮食、饮水或其他过程的影响（如第 1 章示例 1.1 中列文虎克所示）。然而，对于每个人来说，尽管存在每日及每月（月经周期）的微生物量的变化，口腔微生物的组成却出乎意料得稳定[54]。当不存在可发酵的碳水化合物时，链球菌附着在被唾液包裹的牙齿表面；变形链球菌等链球菌发酵糖并产生乳酸，由此促进了在龋齿发展过程中所需的酸性环境[55]（更多疾病相关内容详见第 4 章）。

唾液中的不同组分，来自不同的唾液腺，例如腮腺分泌的唾液含有大量的淀粉酶和较少量的溶菌酶，不含黏蛋白；而下颌下/舌下腺分泌的唾液富含黏蛋白、胱氨酸和溶菌酶，但淀粉酶较少[56]。在其他黏膜部位激活的产生抗体的浆细胞，会再循环到口腔的淋巴组织（如扁桃体），分泌的 IgA 和 IgG 可以和许多口腔细菌结合。咀嚼可诱导牙龈驻留辅助性 T 细胞 17（Th17）的产生[57]。除了来自牙周袋的龈下菌斑外，口腔微生物组大多含有需氧菌或兼性厌氧菌（因此对于细菌来说，在我们睡眠且口腔闭合时，进入肺部是一个好选择[58]）。

我们通过直接观察舌苔完整结构，发现细菌在生长成多层的过程中，可以相对于周围的细菌进行伸缩（图 2.7）[59]。这种局部的动态变化，可能是由细菌的生长速度不同、细胞的脱落或宿主的防御系统的作用所导致的。回到第一章中关于微生物数目的问题，口腔包含了多种不同生境，有必要对固体表面和液体（唾液、龈沟液）中微生物的决定因素作深入研究。除了特定微生物在一个生态位中大幅增多或减少（图 2.8、图 2.9），很多时候微生物邻居们之间的相互推搡维护了相对丰度的长期稳定。

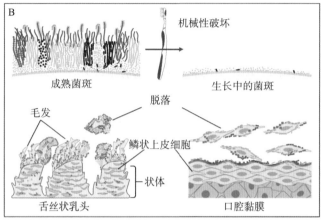

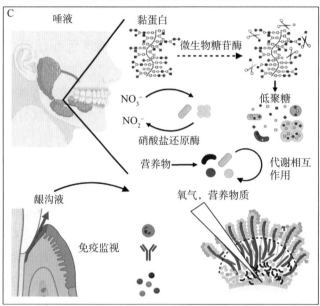

图 2.6　口腔龋洞中微生物面对的选择压

A. 唾液流动和附着。B. 细胞脱落和定植。C. 宿主和微生物群落。宿主和微生物群落之间，会经由附着相互作用、免疫巡查、营养和溶质的梯度变化，产生互利的影响。宿主分泌的唾液黏蛋白，是一种复杂的糖蛋白，能够为多种互养的微生物提供寡糖的来源，而这些微生物则具有分解黏蛋白的糖苷酶。另外，宿主唾液中分泌的硝酸盐和其他营养物质，也能通过龈沟液进入口腔，也可以促进特定微生物的生长，而免疫巡查限制了其他微生物的生长。反过来，微生物代谢可以使局部氧气和营养大量富集。微生物在这些区域中的有利位置的出现，可以影响微生物群落中的代谢相互作用和空间结构。来源：Cell Host Microbe, 2020, 28:160-168. https://doi.org/10.1016/j.chom.2020.07.009.

第 2 章 微生物群落生态

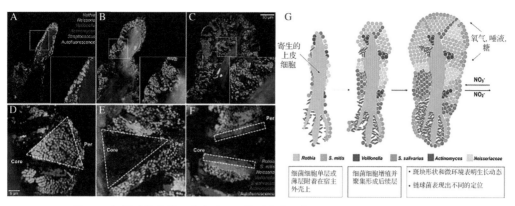

图 2.7 根据舌苔内外层宽度差异，推测出的舌苔微生物生长动态过程的案例

寄生的上皮细胞（自发荧光）被数量不断增加的口腔细菌细胞包围（A~F）。生物膜厚度的变化梯度，以及寄生域的形状，会预示生物膜的生长和选择优势。A. 由少量不同类群的细菌构成的薄生物膜。B. 更厚一些的生物膜，展示出兼性厌氧菌Rothia、厌氧菌Veillonella和放线菌的扩展。C. 一个边界清晰的成熟结构。D. 增加单克隆区域朝向周边的宽度，说明了在边缘有选择优势。E. 朝向边缘的宽度降低，说明了在边缘菌的生长劣势或内部的生长优势。F. 相等的宽度，说明在相邻的分类单元中，不存在优势也不存在劣势。G. 推测出的舌苔微生物群落的发育。该研究关注整体的形态。受试者在监督下，用一个凹凸不平的塑料采样器，从后向前轻柔地刮舌背表面。随着细菌的繁殖，形成多层片状的结构。某些链球菌细胞在表面形成薄膜。边界的形成，是独立于周边及微环境的。细菌可以从宿主的上皮细胞中获取一些营养物质，从口腔唾液中获取一些氧气和硝酸盐。来源：Cell Rep, 2020, 30:4003-4015.e3. https://doi.org/10.1016/j.celrep.2020.02.097.

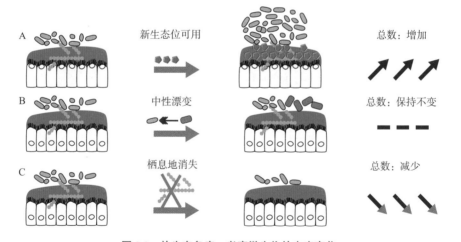

图 2.8 从生态角度，考察微生物的丰度变化

A. 之前不存在的生态位现在存在了。B. 环境的整体承载能力不变，但邻近的微生物的变化或其他来源微生物替换已有微生物，造成了细菌个数动态的改变。C. 由于其他微生物或者宿主的重构，或由于对食物以及寄居地的竞争，导致细菌失去了生存环境。

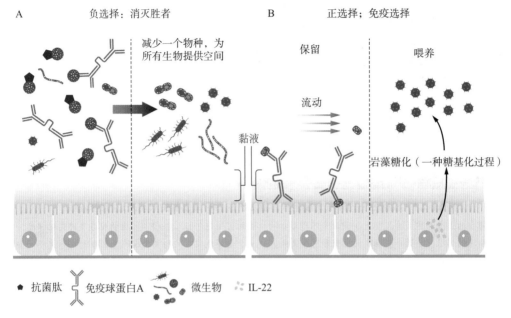

图 2.9 微生物多样性的免疫调控

A. 负向选择：消灭优势者。微生态学研究发现，捕食者可以通过抑制过度适应环境和过度增殖的物种，来增加菌群的多样性。限制这些物种的种群增长，可以解放出生态位，从而使其他物种得以繁衍。免疫系统经由多种机制，以负向选择控制肠道微生物群落。抗菌肽（AMPs）可以介导直接杀伤细菌，IgA 可以引起特定微生物的聚集和清除。B. 正向选择：免疫选择。免疫系统也可以选择特定的微生物在肠道内定居。IgA 除了可以进行负向选择外，还可以通过增加生长缓慢的菌种在黏液中的停留时间，或者使它们在受保护的生态位（如结肠隐窝）中驻留，来保留特定的细菌类群。免疫系统也可以通过诱导特定营养物质的腔内沉积来支持特定类群的存活和生长，例如，白细胞介素-22（IL-22）诱导上皮岩藻糖基化，滋养特定的有益的细菌分类单元。来源：Fig. 3 Sci Immunol, 2018, 3:eaao1603. https://doi.org/10.1126/sciimmunol.aao1603。

2.5　一个稳定的肠道菌群

2.5.1　附着作用

粪便微生物组不是一个原位的群落，我们需要一些想象力和计算模拟来理解没给微生物为什么没有丢失，而是保持了每天的种群稳定。微生物在肠道中的定殖，取决于它们能否黏附在自己喜欢的黏蛋白聚糖上，或者借助邻近微生物的聚糖或其他大分子，或者得到人类宿主的其他帮助。

IgA 是黏膜抗体的一种主要类型，它的作用不仅是通过形成多聚体来清除病原体，还能通过与许多共生细菌形成弱结合，来帮助它们在肠道中的保留[60-63]。如果没有被链球菌蛋白酶切开，IgA 也有可能在口腔中保留[64]。对肠道沙门氏菌鼠伤寒亚种，血清型鼠伤寒沙门菌，高亲和力的 IgA（抗体）能够阻止细胞分裂后的子代细胞分离，这可能是因为 IgA 能够形成多聚体的能力，依赖于相对比例较大的正在分裂的细菌。

IgA 的分泌主要发生在小肠，但是 IgA 及其结合的肠道细菌，包括放线菌门的厚壁菌门和双歧杆菌，都可以从粪便样品中沉淀分离[61,63,66]。例如，大肠埃希菌，在 IgA 缺乏的个体中丰度过多，同时如毛螺菌和瘤胃球菌科的细菌丰度不足[63,67,68]。小肠的微生物群落，主要由乳杆菌科和肠杆菌科细菌组成[69]，小鼠实验表明，小肠 IgA 会与肠杆菌科细菌和其他变形杆菌（如不动杆菌）和双歧杆菌结合[61]。

值得注意的是，IgAs 也是复合聚糖的来源，可以被拟杆菌等细菌切割和代谢[70]，而聚糖的变化可能会影响 IgA 与其他细菌的结合。

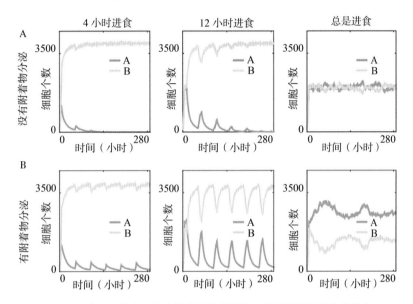

图 2.10　流动环境中，宿主的附着性分泌物，能够促进微生态的稳定

宿主的附着性分泌物，可以防止特定区域寄生的微生物被冲走。McLoughlin等模拟了两种基因型，一个通用型的物种B，其以常见的流动营养物为食。另一种是专家型的物种A，它只在特定时刻出现的营养物为食。我们展示了在每48小时的周期中，每4小时和12小时断食的情况下，与第二种营养物质可以获得的情况进行对比。如果没有宿主的附着性分泌物，物种A在断食周期少且短的时候，就不会存活。当宿主分泌促进附着的因子时（图B），会形成一个水平区域，在这个区域内物种A比物种B能够更难被替代，而两种物种都能够在环境变化下存活。在始终喂食的环境中，种群的变化，来源于种群大小的随机波动。因此该研究的结论为，在不同的断食周期和持续时间下，都是稳健的。来源：Cell Host Microbe, 2016, 1-10. https://doi.org/10.1016/j.chom.2016.02.021.

2.5.2　肠道的蠕动

蠕动混合不仅决定了食物残渣通过肠道的所需时间，还确保了一部分微生物能够在相邻处被挤压（图 2.11、图 2.12）。人类的一个特点是，当我们直立时，升结肠实际上是垂直的。根据模拟（图 2.12），升结肠是食物残渣从小肠进入结肠时主要的生物反应器[71]。很多动物，包括小鼠，都有一个很大的盲肠来储存微生物。据推测，人类阑尾中的微生物可能也有助于在遭受严重干扰后，恢复微生物群落的平衡[72]。

然而，一些微生物可能来自更远的消化道，或局部存在于结肠黏膜（图2.11C、第1章，图1.10）[73]。淋巴系统可能携带一些来自其他身体部位的微生物。食物残渣在升结肠中用黏蛋白包裹，在结肠末端则被更多的黏蛋白包裹[74]。上皮细胞和管腔之间存在细微的差异。脆弱拟杆菌在黏蛋白或组织中比在管腔中表达更多的硫酸酯酶和糖苷水解酶。而大肠埃希菌在铁的获取途径上有所不同[75]。

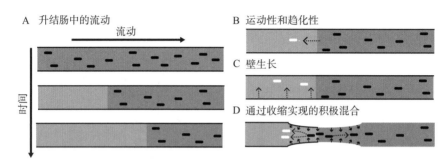

图2.11 肠道蠕动造成的冲刷和可能的抵消因素

在成年人的近端结肠内，管腔内包含的物质以约20 μm/s的平均速度沿结肠移动。A. 随着时间的推移，仅肠道蠕动就会导致管腔的排空。因此需要其他因素来抵消肠道蠕动的影响，并随着时间的推移帮助保持细菌密度的稳定。这些因素可能包括：B. 细菌的主动运动，向着营养源游动；C. 细胞壁生长；D. 蠕动混合，使管腔内的物质与结肠壁的反流物质相混合。来源：https://doi.org/10.1073/pnas.1601306113的图1.

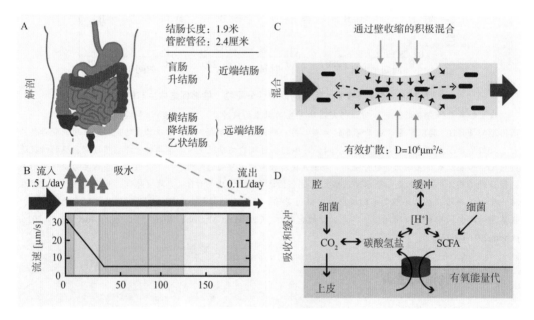

图2.12 人类结肠的生理参数

A. 解剖维度。Cremer等根据尸检时对人体结肠解剖的测量，以及使用造影剂的X射线、CT成像和磁共振断层成像等技术，推导出不同结肠段的长度、表面积和管腔直径。B. 管腔流量。每天大约有1.5 L的液体到达近端结肠。上皮细胞吸收了其中的大部分，每天只有100~200 mL以粪便的形式排出结肠。这种沿着结肠的液体吸收持续进行，腔流速度也随之发生了显著的变化。我们估算了结肠起始处的平均流速约为30 μm/秒，而到了升结肠末端则下降到约5 μm/s。C. 混合腔内容物。肠壁的收缩可以产生局部混合。我们根据放射性标记染料在大肠内混合的数据，推导出所测得的分布可以用一个有效扩散常数——d-10⁶ μm²/s来近似，这个数值比分子扩散的速率高了一个数量级。D. 上皮细胞对短链脂肪酸SCFA的摄取、碳酸氢盐排泄和缓冲化合物。细菌发酵产生的短链脂肪酸被肠道上皮细胞吸收，为宿主提供了能量来源。短链脂肪酸的摄取与碳酸氢盐（bicarbonate）的排泄相耦合，而碳酸氢盐与CO_2和其他管腔组分处于平衡态，可缓冲管腔酸度。所有的计算都基于观测到的上皮细胞转运蛋白的特性和腔的缓冲能力。来源：Proc Natl Acad Sci U S A, 2017, 114:6438–43. https://doi.org/10.1073/pnas.1619598114.

根据标准西方饮食进行模拟，微生物可能会按照一定的顺序沿着近端结肠富集分布，拟杆菌先于厚壁菌[71]。甲烷主要在远端结肠产生（示例2.3）[76]。如果饮食中含有更多的纤维，发酵过程可以从升结肠一直持续到直肠（第8章，表8.4）[77]。有趣的是，标准饲料喂养（chow diet，而不是高脂饮食）的小鼠，即使不限制食物供应的时间，拟杆菌、厚壁菌、疣微菌门等肠道菌也能形成昼夜节律（示例2.3）[78]。虽然根据宏基因组数据开展的微生物生长率分析在早产儿、枸橼酸杆菌感染、炎症性肠病（IBD）和糖尿病患者的数据中发现了与正常样本的差异[79-81]，但作者倾向于认为，在健康成年人中，大部分粪便中的微生物群落，在到达直肠时已经不再显著增殖。厚壁菌（与消耗氢气的产甲烷菌一起）也许更容易利用每次进食的机会快速生长（图2.13）[78,82]。对于微生物来说，传统的农业利用人类和动物的粪便的做法，可能是一个更加有意义的循环，现代人每天排出的粪便不足100 g[83]（示例2.3）。

示例2.3　和人类粪便微生物有显著关联的人类基因

乳糖代谢的 LCT 基因

乳糖不耐受（lactose-resistant，LCT）的个人在摄入乳糖（如牛奶）后可能会有更多的双歧杆菌属。乳糖不耐和双歧杆菌属之间的全基因组关联分析，在众多欧洲人群中的信号最强，中国人几乎都乳糖不耐，在以色列样本中较弱[42,125]。厚壁菌门中，*Negativibacillus* 与一种瘤胃球菌（*Ruminococcus* sp. UBA3855），也和乳糖不耐的 LCT 基因位点相关[126]。在血清水平，色氨酸代谢产物吲哚丙酸，可能经由双歧杆菌属，与乳糖不耐的 LCT 基因位点相关。吲哚丙酸和膳食纤维与2型糖尿病的风险降低有关，在同一研究中，膳食纤维和厚壁菌门中的丁里弧菌、反刍球菌、真菌和细菌纤维素单胞菌正相关[127]。

从古至今，很多动物，包括骆驼、马都是奶制品的来源，每一样都有不同的聚糖结构和丰富程度[128,129]。根据考古学证据，早在6000年前的新石器时代，就有记载的乳制品摄入行为，而这比可代谢乳糖的 LCT 基因型出现的时间早了2000年[130]。LCT 基因编码的乳糖酶（从山羊小肠中提纯）已被证明能水解在植物中常见的黄酮类化合物和异黄酮[131]。能代谢乳糖的 LCT 基因型在人群中的传播，目前多被归因于游牧人群[128]。

ABO 血型影响

粪便微生物和 ABO 基因的相关性,在欧洲人的样本中同样显著[126,132]。ABO 组织血型和 FUT2 分泌状态,与拟杆菌和粪便杆菌的丰度相关,而这对炎症性肠炎等病的发生有影响[132]。不同人群中的 ABO 血型的分布可以有显著差异。例如,在疟疾泛滥的区域,O 型血能占据 50% 以上的人群[133,134]。B 型血对霍乱弧菌感染风险更低[135]。造成胃癌的幽门螺杆菌,可能会和不同的 ABO 血型群以不同方式结合[133,136]。

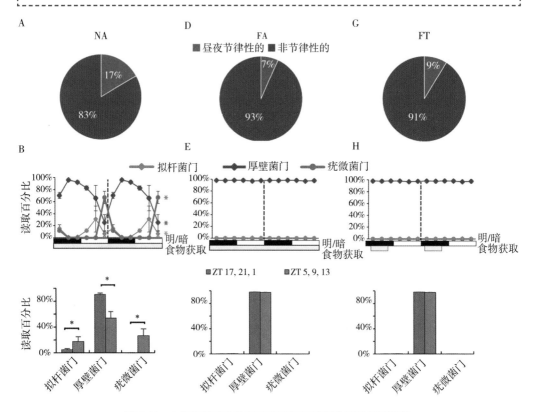

图 2.13 小鼠肠道微生物群落在不同饲养条件下的昼夜节律

NA 小鼠可以自由进食正常的食物。FA 小鼠可自由进食高脂饮食 HFD。FT 小鼠只能在 8 h 内(ZT 13~21)进食 HFD。A. 饼图显示 NA 小鼠($n=18$)处于昼夜节律性和非节律性的 OTUs(在所有情况下)所占的百分比。B. 上双曲线图-其中第二个循环是第一个循环的重复,显示了每个时间点($n=3$)三个最主要的门的平均读数比例($\pm$SEM)。底部黑色和白色的框分别表示关灯和开灯的时间。黄色的框显示老鼠可以获得食物的时间。根据 JTK 分析结果,线性图中线末端的彩色星号表示相应门的丰度具有昼夜节律(即 ADJ.$P<0.05$ 和 BHQ<0.05)。由于食物丸到达盲肠需要超过 1 h[84],较低的柱状图显示了黑暗/活跃喂养阶段(ZT 17,21,1)和光照/禁食阶段(ZT 5,9,13)的平均读数($\pm$SEM,$n=9$)。*$P<0.05$。C. 极坐标图描绘了排名前十的 OTU(基于读数百分比)。弧度表示 OTU 峰值的相位,距离中心是所有时间点的平均读数百分比,每个点的半径表示循环的振幅。圆圈的颜色表明 OTU 的门:厚壁菌(粉红色),拟杆菌(蓝色),疣微菌门(绿色)。图中左侧的黑色弧线表示光明/黑暗的循环。黄色的弧线表示可以进食的时间。底部极坐标图显示了放大的内环(10%)的顶部极坐标

图。这些描述也适用于FA鼠（D、E、F）和FT鼠（G、H、I）的子图。来源：Cell Metab, 2014, 20:1006-1017. https://doi.org/10.1016/j.cmet.2014.11.008.

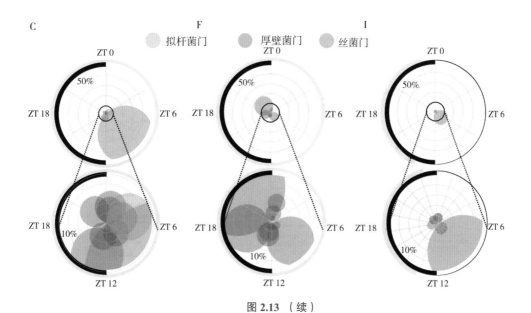

图2.13（续）

图2.14 肠道微生物的丰度在健康人和患动脉粥样硬化的患者中，呈现出协同变化模式

黄色虚线圆圈显示拟杆菌属和普雷沃菌属。紫色的圆圈突出了厚壁菌门的成员，它们可能清除拟杆菌属等产生的聚糖在内的代谢物。A. 187个健康对照组的物种间的相关性。根据Wilcoxon秩和检验$q<0.05$，对照组和急性脑血管病（ACVD）患者的相对丰度（青色圈代表控制组富集的基因红色圆圈代表ACVD富集的基因）显示圆圈有颜色。绿线，Spearman相关系数>0.3；红线，Spearman相关系数<-0.3。B. 218例急性脑血管病患

者的物种间的相关性。其中205例患者有稳定型心绞痛，8例有不稳定型心绞痛，5例有心肌梗死。来源：Nat Commun, 2017, 8:845. https://doi.org/10.1038/s41467-017-00900-1.

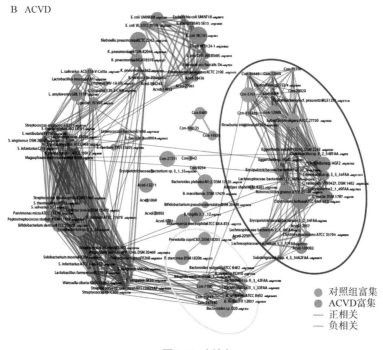

图2.14 （续）

2.6 肠型和塞伦盖蒂法则

肠型，这是一个有争议的概念，本质上是对粪便微生物数据进行非监督聚类的结果。统计上最优的聚类个数，取决于样本来源的队列[85]。对于居住在城市的东亚人群，以及更少被研究的非洲的农业人群，他们的肠道中厚壁菌（包括不胖有关的克里斯滕林氏菌）和产甲烷古菌（示例2.4）（史密斯甲烷菌，图2.2）发酵链似乎没有北欧人群那么丰富[86-88]，结肠的pH也没有那么低[69,71]。梭状芽孢杆菌能够抑制小鼠肠道内维生素A的合成，可能在断奶后，完成对调节性T细胞的诱导，而成年人的粪便梭状芽孢杆菌与血浆维生素A有关[36,41,89,90]。但是目前还没有研究证明维生素A水平的人群差异与肠道微生物组成有关。厚壁菌的数量也与进食周期有关（图2.13）。根据 *gyrB* 基因（DNA旋转酶b亚单位）分析肠道菌与宿主的共进化，人类、黑猩猩和大猩猩比较，发现在拟杆菌科中，这种共生模式最为显著，其次是双歧杆菌科（示例2.4），在毛螺菌科（Lachnospiraceae）中，这种共生模式最为微弱[91]。这些结果也与厚壁菌的孢子形成能力一致，相对于普氏杆菌、拟杆菌和双歧杆菌，厚壁菌更容易在个体之间传播。

第 2 章 微生物群落生态

> **示例 2.4 人类菌群中的产甲烷古菌（代谢氢和 TMAO）**
>
> 史密斯甲烷短杆菌（*Methanobrevibacter smithii*）是一种能够利用肠道中发酵产生的氢气，将其转化为甲烷的微生物（例如图 2.2 中的克里斯滕森氏菌产生的氢气）。它在欧洲人的肠道中比东亚人的肠道中丰度明显更高[86]。甲烷的产生主要发生在结肠末端，与炎症性肠病、结直肠癌等有关[76]。
>
> 在反刍动物和其他动物的消化道中也发现甲烷短杆菌[119,137]。来自牛瘤胃的瘤胃甲烷短杆菌（*M. ruminantium*）M1 菌株编码了一种黏附素，可以黏附产氢微生物，如瘤胃原生动物（包括 *Epidinium caudatum* 和 *Entodinium* spp.）和属于细菌的蛋白溶解丁酸弧菌（*Butyrivibrio proteoclasticus*）[138]。
>
> 一种不如史密斯甲烷短杆菌有名的产甲烷菌——卢米尼甲烷马赛球菌，已被证明能够利用氢气来还原包括三甲胺（TMA）等甲基化合物。这种分解 TMA 的能力[139]，可能有助于减轻细菌性阴道炎（BV）患者的鱼腥味（也可能源于尸胺或腐胺[140]），或防止氧化三甲胺（trimetlylamine oxide，TMAO）引发的动脉粥样硬化[141,142]。然而，TMA 和 TMAO 在海鱼中含量较高[143]；TMA 是从尿液和汗液中排出的[144]；给成年雄性小鼠喂食间接 TMAO 抑制剂后，它们在社交竞争中的胜率降低，与其社会地位无关[145]，这意味着 TMA 在繁殖期具有进化优势。在小鼠体内，只在雄性的肝脏中才有三甲胺氧化酶，它是一种含有黄素的单加氧酶 3（FMO3），能够氧化三甲胺，而三甲胺则被嗅觉受体——微量胺相关受体 5（TAAR5）识别，作为一种物种特异性的吸引信号[146,147]。

设想一下，用生活在塞伦盖蒂大草原的动物进行类比[94]，丰度最高的物种，位于生态系统的最底层，像采集阳光一样决定着生态系统的能量输入（图 2.1）。在东亚人群中，普雷沃菌或拟杆菌占据主导地位，会呈现出显著的分布模式[95]（而非密螺旋体，示例 2.2）。与美国和欧洲的粪便样本相比，来自墨西哥当代和 1000 年前（粪化石）的样本中，粪便普雷沃菌的含量更高[96,97]。相比于普雷沃菌占主导的人群，在拟杆菌占主导的人群中，很多成熟的疫苗有效性更强[98]。在实验室的小鼠中，除了专门选择普雷沃菌占主导的情况外，也是由拟杆菌和厚壁门菌占主导[99,100]。那些拟杆菌相对丰度更高的人，通常没有寄生虫和原虫的感染[101-103]。人类的基因差异，也会影响普雷沃菌和拟杆菌的丰度差异[42,104]。相比于从太阳中获取能量并构建食物链，人体微生物不仅仅是待在那里或是化能自养，还能从宿主的分子，或者从食物、饮料和药物中获得能量。拟杆菌属，特别是多形拟杆菌，能够在细菌细胞外，消化来自黏蛋白、IgA 或食物中的复杂多糖[70,105-108]。

这些被分解的聚糖在进入细胞内进一步代谢之前，可能会被其他细菌所利用，其中包括多种厚壁菌。一项将 15 种菌混合移植到无菌小鼠肠道的研究，通过荧光探针标记发现，拟杆菌属的几种菌待在一起。采用 19 μm 见方的网格分析，解纤维素拟杆菌（*B. cellulolyticus*）与普通拟杆菌（*B. vulgatus*）的数量呈正相关，而解纤维素拟杆菌与扭链瘤胃球菌（*Ruminococcus torques*）的数量呈负相关。[109]。菊粉的胞外消化能够提高卵形拟杆菌的适应性，因为它可以与其他肠道物种，如普通拟杆菌[108]，形成互利关系。由于技术局限性，目前我们还无法研究人类肠道微生物中这种局部合作关系。昆虫的简单种群，为我们提供了一个很好的范例，展示了基因差异和代谢的相互作用，是如何塑造一个种群的空间结构的（图 2.15）。

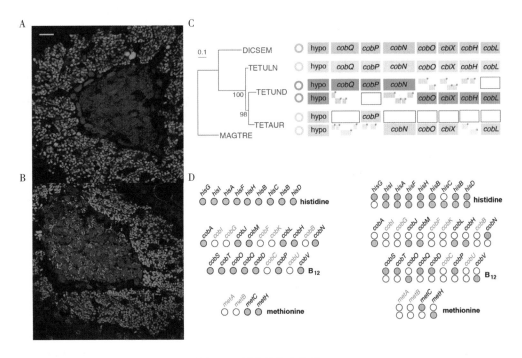

图 2.15　原位的共生关系

通过了解微生物群落的空间组成和基因（代谢潜能）的信息，研究得到 TETUND1 和 TETUND2 之间代谢层面的关系，这两种细菌都是细胞内共生菌，寄生在叶蝉属的蝉昆虫体内。A. 利用荧光原位杂交技术（FISH）检测 RNA，以区分霍奇金氏病 Hodgkinia（红色）和内共生菌（绿色）。B. 利用荧光原位杂交技术检测基因组，以区分 Sulcia（绿色）、TETUND1（黄色）和 TETUND2（蓝色）三种细菌。C. 和 D. TETUND1（橙色）和 TETUND2（蓝色）的基因组分化出互补的功能。半环蝉（*Diceroprocta secincta*，DICSEM）仅有沟纹和绿色霍奇金蝉（Hodgkinia）来源：Cell, 2014, 158:1270-1280. https://doi.org/10.1016/j.cell.2014.07.047.Fig. 2.14.

思考题 2.2

（1）你认为什么时候我们能够在样本中，直接观察到所有微生物并获得其完整的基因组和转录组信息呢？

（2）如果我们拥有了这些信息，又能解决那些问题？

2.7 总结

本章从宏观生态学的角度，引入了物种多样性和营养级的概念。在某些疾病状态下，微生物多样性或丰富性可能会增加，尤其是在考虑肠道以外的微生物组时。在绝经后的大多数妇女中，单种阴道菌群的优势地位变得更加多样化（第8章会涉及），因为不再维持酸性水平（由乳酸杆菌发酵乳酸糖原而产生）。在皮脂腺分泌随着年龄的增长而减少之前，皮肤的脂类、微量元素和其他成分为皮肤提供了稳定的偏酸性的微生物群落。空间分离和非特异性交叉互养是维持微生物群落稳定性的关键因素。黏蛋白、免疫球蛋白，甚至是作为血型抗原表达的聚糖，都能帮助微生物定位，并提供营养。对于肠道微生物组而言，主要的生物反应器是升结肠，而微生物可能沿着结肠扩散，发酵它们能够利用的物质。在婴儿和老年阶段，微生物组会主动和被动地获得更多的成员。人类基因差异、历史偶然性和昼夜节律（示例2.5），以及被广泛研究的因素，如营养物质和免疫反应，都对目前在微生物组的组成有影响。正常菌群的营养级可能需要考虑不同的生理状态，因为在不同的状态下，可利用的分子不同，从而导致不同的功能具有不同的进化优先级。

示例 2.5　肠道微生物的昼夜节律及其产物

无菌小鼠会因回肠皮质酮分泌过量，长期处于糖尿病前期的状态[148]。核受体 RORα 激活剂和 REVERBα 抑制剂共同在肠上皮表达 TLRs（Toll 样受体），进而产生昼夜节律。微生物通过 TLRs 信号诱导 JNK 和 IKKβ，抑制 PPARα 并激活 REVERBα，形成节律性活动。最近已有研究在 2 型糖尿病和肥胖患者中检出节律异常的肠道微生物组[149]。

如果图 2.13 中的疣微菌门，确实是能够降解黏蛋白的阿克曼氏菌（Akkermansia），那么它们数量增长的一个可能的原因是，在一天的结束时（小鼠昼伏夜出），当食物中的营养被消耗完后，肠道微生物可依赖宿主的黏蛋白生活。这种依赖黏蛋白的存续，在下一次进食前，不足以维持一个足够大的厚壁菌的种群规模[78,149,150]。丁酸盐在睡眠后很快达到峰值[151]。产乙酸盐或丙酸盐的嗜黏蛋白阿克曼氏菌（Akkermansia muciniphila）增加了生热作用，但随着温度降低而减少[152,153]。这种细菌很好地适应了肠道内壁松散的外层黏液层，并且在二氧化碳存在的情况下能够利用纳摩尔浓度的氧气[154]。人体内的氧气水平也呈现昼夜节律[122,155]。在患有阿尔茨海默病、冠状动脉疾病和精神分裂症的人群中，嗜黏蛋白阿克曼氏杆菌的相对丰度较对照人群高[25,156-158]。

粪便样本的采样，通常是在早上进行，可能是在早餐前后。对于人类受试者来说，在一天中的不同时间点采集粪便样本，存在技术上的难度[159]。前面提到的2型糖尿病的研究，仔细地记录了排便的时间，而这解释了部分肠道微生物的粪便中的一些微生物，包括阿克曼菌（*Akkermansia*）、罗氏菌属（*Roseburia*）、长双歧杆菌、粪便杆菌，在2型糖尿病患者中，展现了不同的节律震荡。

小肠上皮细胞中的组蛋白脱乙酰酶3（HDAC3）有节律性表达[160]。HDAC3促进了营养物质转运和脂质代谢相关基因的表达，并激活了雌激素相关受体α，从而促进脂质吸收。小肠上皮细胞的组蛋白甲基化也呈现出节律性[78]。

多胺（腐胺、精胺和精胺）有助于维持正常的生物钟，随着年龄的增长，多胺水平下降，周期延长[161]。缺乏多胺的饮食会影响肝脏转录组的昼夜振荡[78]。此外，喂食高脂肪饲料的小鼠在进食后会从dsrAB（异化亚硫酸盐还原酶）中产生亚硫酸氢盐，这将在肠道中产生炎症[151]。

肺微生物组也可能存在昼夜节律的，更多的微生物在夜间从口腔迁出，白天减缓迁入，同时在白天会有更多的细菌被清除[58]。

有些微生物自身有调控节律的基因。一些蓝细菌有完整的*KaiA*、*KaiB*、*KaiC*基因，或只有*KaiB*和*KaiC*，或只有*KaiC*基因的蓝细菌，也具有维持节律的能力，这已在一些变形杆菌和蓝细菌中被发现[162]。肠道、泌尿道和呼吸道的细菌/变形菌产气菌种克雷伯菌（以前属于肠杆菌属）在体外有一个聚集和运动的昼夜模式，而在加入褪黑素后，昼夜模式的稳定性增强[163,164]。此外，气温也会影响着昼夜节律。这说明，微生物可从宿主接收有关昼夜节律的潜在模式。枯草杆菌不编码*KaiA*、*KaiB*、*KaiC*，但有光感受器和Per-Arnt-Sim（PAS）域；枯草芽孢杆菌呈现出一个持续24 h，受到光或温度控制的节律[165]。

思考题 2.3

请你尝试描述正常环境下菌群的食物链。例如，富含膳食纤维的肠道微生物组，类似于牛的瘤胃，在夜晚饥饿时，肠道微生物组食用黏蛋白和其他潜在的宿主分子。

原著参考文献

[1] Meyer ST, Ptacnik R, Hillebrand H, Bessler H, Buchmann N, Ebeling A, et al. Biodiversity–multifunctionality relationships depend on identity and number of measured functions. Nat Ecol Evol 2018;2:44–49. https://doi.org/10.1038/s41559-017-0391-4.

[2] Martiny JBH, Jones SE, Lennon JT, Martiny AC. Microbiomes in light of traits: a phylogenetic perspective. Science 2015;350:aac9323. https://doi.org/10.1126/science.aac9323.

[3] Goldford JE, Lu N, Bajić D, Estrela S, Tikhonov M, Sanchez-Gorostiaga A, et al. Emergent simplicity in microbial community assembly. Science 2018;361:469–474. https://doi.org/10.1126/science.aat1168.

[4] Coyte KZ, Schluter J, Foster KR. The ecology of the microbiome: networks, competition, and stability. Science 2015;350:663–666. https://doi.org/10.1126/science.aad2602.

[5] Lourenço M, Chaffringeon L, Lamy-Besnier Q, Pédron T, Campagne P, Eberl C, et al. The spatial heterogeneity of the gut limits predation and fosters coexistence of bacteria and bacteriophages. Cell Host Microbe 2020;28:390–401.e5. https://doi.org/10.1016/j.chom.2020.06.002.

[6] Costello EK, Stagaman K, Dethlefsen L, Bohannan BJM, Relman DA. The application of ecological theory toward an understanding of the human microbiome. Science 2012;336:1255–1262. https://doi.org/10.1126/science.1224203.

[7] Song SJ, Lauber C, Costello EK, Lozupone CA, Humphrey G, Berg-Lyons D, et al. Cohabiting family members share microbiota with one another and with their dogs. Elife 2013;2. https://doi.org/10.7554/eLife.00458, 458.

[8] van Leeuwenhoek A. The collected letters of Antoni van Leeuwenhoek. Amsterdam: Swets and Zeitlinger; 1952.

[9] van Leeuwenhoek A. Observations, communicated to the publisher by Mr. Antony van Leewenhoeck, in a dutch letter of the 9th Octob. 1676. here English'd: concerning little animals by him observed in rain-well-sea- and snow water; as also in water wherein pepper had lain infus. Philos Trans R Soc London 1677;12:821–831. https://doi.org/10.1098/rstl.1677.0003.

[10] Yildirim S, Yeoman CJ, Janga SC, Thomas SM, Ho M, Leigh SR, et al. Primate vaginal microbiomes exhibit species specificity without universal Lactobacillus dominance. ISME J 2014;8:2431–2444. https://doi.org/10.1038/ismej.2014.90.

[11] Gliniewicz K, Schneider GM, Ridenhour BJ, Williams CJ, Song Y, Farage MA, et al. Comparison of the vaginal microbiomes of premenopausal and postmenopausal women. Front Microbiol 2019;10:193. https://doi.org/10.3389/fmicb.2019.00193.

[12] Jie Z, Chen C, Hao L, Li F, Song L, Zhang X, et al. Life history recorded in the vagino-cervical microbiome along with multi-omics. Genomics Proteomics Bioinformatics 2021. https://doi.org/10.1016/j.gpb.2021.01.005.

[13] Qi W, Li H, Wang C, Li H, Fan A, Han C, et al. The effect of pathophysiological changes in the vaginal milieu on the signs and symptoms of genitourinary syndrome of menopause (GSM). Menopause 2021;28:102–108. https://doi.org/10.1097/GME.0000000000001644.

[14] Turnbaugh PJ, Hamady M, Yatsunenko T, Cantarel BL, Duncan A, Ley RE, et al. A core gut microbiome in obese and lean twins. Nature 2009;457:480–484. https://doi.org/10.1038/nature07540.

[15] Cotillard A, Kennedy SP, Kong LC, Prifti E, Pons N, Le Chatelier E, et al. Dietary intervention impact on gut microbial gene richness. Nature 2013;500:585–588. https://doi.org/10.1038/nature12480.

[16] Le Chatelier E, Nielsen T, Qin J, Prifti E, Hildebrand F, Falony G, et al. Richness of human

gut microbiome correlates with metabolic markers. Nature 2013;500:541–546. https://doi.org/10.1038/nature12506.

[17] He Q, Gao Y, Jie Z, Yu X, Laursen JMJM, Xiao L, et al. Two distinct metacommunities characterize the gut microbiota in Crohn's disease patients. Gigascience 2017;6:1–11. https://doi.org/10.1093/gigascience/gix050.

[18] Ricanek P, Lothe SM, Frye SA, Rydning A, Vatn MH, Tønjum T. Gut bacterial profile in patients newly diagnosed with treatment-naïve Crohn's disease. Clin Exp Gastroenterol 2012;5:173–86. https://doi.org/10.2147/CEG.S33858.

[19] Gevers D, Kugathasan S, Denson LA, Vázquez-Baeza Y, Van Treuren W, Ren B, et al. The treatment-naive microbiome in new-onset Crohn's disease. Cell Host Microbe 2014;15:382–92. https://doi.org/10.1016/j.chom.2014.02.005.

[20] Imhann F, Vich Vila A, Bonder MJ, Fu J, Gevers D, Visschedijk MC, et al. Interplay of host genetics and gut microbiota underlying the onset and clinical presentation of inflammatory bowel disease. Gut 2018;67:108–19. https://doi.org/10.1136/gutjnl-2016-312135.

[21] Manichanh C, Rigottier-Gois L, Bonnaud E, Gloux K, Pelletier E, Frangeul L, et al. Reduced diversity of faecal microbiota in Crohn's disease revealed by a metagenomic approach. Gut 2006;55:205–11. https://doi.org/10.1136/gut.2005.073817.

[22] Maldonado-Arriaga B, Sandoval-Jiménez S, Rodríguez-Silverio J, Lizeth Alcaráz-Estrada S, Cortés-Espinosa T, Pérez-Cabeza de Vaca R, et al. Gut dysbiosis and clinical phases of pancolitis in patients with ulcerative colitis. Microbiology Open 2021;10. https://doi.org/10.1002/mbo3.1181, e1181.

[23] Feng Q, Liang S, Jia H, Stadlmayr A, Tang L, Lan Z, et al. Gut microbiome development along the colorectal adenoma–carcinoma sequence. Nat Commun 2015;6:6528. https://doi.org/10.1038/ncomms7528.

[24] Sanapareddy N, Legge RM, Jovov B, McCoy A, Burcal L, Araujo-Perez F, et al. Increased rectal microbial richness is associated with the presence of colorectal adenomas in humans. ISME J 2012;6:1858–68. https://doi.org/10.1038/ismej.2012.43.

[25] Zhu F, Ju Y, Wang W, Wang Q, Guo R, Ma Q, et al. Metagenome-wide association of gut microbiome features for schizophrenia. Nat Commun 2020;11:1612. https://doi.org/10.1038/s41467-020-15457-9.

[26] Yang J, Zheng P, Li Y, Wu J, Tan X, Zhou J, et al. Landscapes of bacterial and metabolic signatures and their interaction in major depressive disorders. Sci Adv 2020;6. https://doi.org/10.1126/sciadv.aba8555, eaba8555.

[27] Zhu J, Liao M, Yao Z, Liang W, Li Q, Liu J, et al. Breast cancer in postmenopausal women is associated with an altered gut metagenome. Microbiome 2018;6:136. https://doi.org/10.1186/s40168-018-0515-3.

[28] Moossavi S, Sepehri S, Robertson B, Bode L, Goruk S, Field CJ, et al. Composition and variation of the human milk microbiota are influenced by maternal and early-life factors. Cell Host Microbe 2019;25:324–335.e4. https://doi.org/10.1016/j.chom.2019.01.011.

[29] Kalaora S, Nagler A, Nejman D, Alon M, Barbolin C, Barnea E, et al. Identification of bacteria-derived HLA-bound peptides in melanoma. Nature 2021;592:138–43. https://doi.org/10.1038/

s41586-021-03368-8.

[30] Lamont RJ, Koo H, Hajishengallis G. The oral microbiota: dynamic communities and host interactions. Nat Rev Microbiol 2018;16:745–59. https://doi.org/10.1038/s41579-018-0089-x.

[31] Ravel J, Gajer P, Abdo Z, Schneider GM, Koenig SSK, Mcculle SL, et al. Vaginal microbiome of reproductive-age women. Proc Natl Acad Sci U S A 2010;108:4680–7. http://www.pnas.org/cgi/doi/10.1073/pnas.1002611107.

[32] DiGiulio DB, Callahan BJ, McMurdie PJ, Costello EK, Lyell DJ, Robaczewska A, et al. Temporal and spatial variation of the human microbiota during pregnancy. Proc Natl Acad Sci U S A 2015;112:11060–5. https://doi.org/10.1073/pnas.1502875112.

[33] Lundy SD, Sangwan N, Parekh NV, Selvam MKP, Gupta S, McCaffrey P, et al. Functional and taxonomic dysbiosis of the gut, urine, and semen microbiomes in male infertility. Eur Urol 2021;79:826–36. https://doi.org/10.1016/j.eururo.2021.01.014.

[34] Yu J, Feng Q, Wong SHSH, Zhang D, Liang QY, Qin Y, et al. Metagenomic analysis of faecal microbiome as a tool towards targeted non-invasive biomarkers for colorectal cancer. Gut 2017;66:70–8. https://doi.org/10.1136/gutjnl-2015-309800.

[35] Bäckhed F, Roswall J, Peng Y, Feng Q, Jia H, Kovatcheva-Datchary P, et al. Dynamics and stabilization of the human gut microbiome during the first year of life. Cell Host Microbe 2015;17:690–703. https://doi.org/10.1016/j.chom.2015.04.004.

[36] Al Nabhani Z, Dulauroy S, Marques R, Cousu C, Al Bounny S, Déjardin F, et al. A weaning reaction to microbiota is required for resistance to immunopathologies in the adult. Immunity 2019;50:1276–1288.e5. https://doi.org/10.1016/j.immuni.2019.02.014.

[37] Knoop KA, Gustafsson JK, McDonald KG, Kulkarni DH, Coughlin PE, McCrate S, et al. Microbial antigen encounter during a preweaning interval is critical for tolerance to gut bacteria. Sci Immunol 2017;2. https://doi.org/10.1126/sciimmunol.aao1314, eaao1314.

[38] Wilmanski T, Diener C, Rappaport N, Patwardhan S, Wiedrick J, Lapidus J, et al. Gut microbiome pattern reflects healthy ageing and predicts survival in humans. Nat Metab 2021;3:274–86. https://doi.org/10.1038/s42255-021-00348-0.

[39] Biagi E, Franceschi C, Rampelli S, Severgnini M, Ostan R, Turroni S, et al. Gut microbiota and extreme longevity. Curr Biol 2016;26:1480–5. https://doi.org/10.1016/j.cub.2016.04.016.

[40] Zhang X, Zhong H, Li Y, Shi Z, Ren H, Zhang Z, et al. Sex- and age-related trajectories of the adult human gut microbiota shared across populations of different ethnicities. Nat Aging 2021;1:87–100. https://doi.org/10.1038/s43587-020-00014-2.

[41] Jie Z, Liang S, Ding Q, Tang S, Wang D, Zhong H, et al. A multi-omic cohort as a reference point for promoting a healthy gut microbiome. Med Microecol 2021. https://doi.org/10.1101/585893.

[42] Liu X, Tang S, Zhong H, Tong X, Jie Z, Ding Q, et al. A genome-wide association study for gut metagenome in Chinese adults illuminates complex diseases. Cell Discov 2021;7:9. https://doi.org/10.1038/s41421-020-00239-w.

[43] Man MQ, Xin SJ, Song SP, Cho SY, Zhang XJ, Tu CX, et al. Variation of skin surface pH, sebum content and stratum corneum hydration with age and gender in a large Chinese population. Skin Pharmacol Physiol 2009;22:190–9. https://doi.org/10.1159/000231524.

[44] Luebberding S, Krueger N, Kerscher M. Age-related changes in skin barrier function -

quantitative evaluation of 150 female subjects. Int J Cosmet Sci 2013;35:183–90. https://doi.org/10.1111/ics.12024.

[45] Chen C, Song X, Wei W, Zhong H, Dai J, Lan Z, et al. The microbiota continuum along the female reproductive tract and its relation to uterine-related diseases. Nat Commun 2017;8:875. https://doi.org/10.1038/s41467-017-00901-0.

[46] Leheste JR, Ruvolo KE, Chrostowski JE, Rivera K, Husko C, Miceli A, et al. Pacnes-driven disease pathology: current knowledge and future directions. Front Cell Infect Microbiol 2017;7:81. https://doi.org/10.3389/fcimb.2017.00081.

[47] Javurek AB, Spollen WG, Ali AMM, Johnson SA, Lubahn DB, Bivens NJ, et al. Discovery of a novel seminal fluid microbiome and influence of estrogen receptor alpha genetic status. Sci Rep 2016;6:23027. https://doi.org/10.1038/srep23027.

[48] Byrd AL, Belkaid Y, Segre JA. The human skin microbiome. Nat Rev Microbiol 2018;16:143–155. https://doi.org/10.1038/nrmicro.2017.157.

[49] The Human Microbiome Project Consortium. Structure, function and diversity of the healthy human microbiome. Nature 2012;486:207–214. https://doi.org/10.1038/nature11234.

[50] Oh J, Byrd AL, Park M, Kong HH, Segre JA. Temporal stability of the human skin microbiome. Cell 2016;165:854–866. https://doi.org/10.1016/j.cell.2016.04.008.

[51] Sloan WT, Lunn M, Woodcock S, Head IM, Nee S, Curtis TP. Quantifying the roles of immigration and chance in shaping prokaryote community structure. Environ Microbiol 2006;8:732–740. https://doi.org/10.1111/j.1462-2920.2005.00956.x.

[52] Li L, Ma Z. Testing the neutral theory of biodiversity with human microbiome datasets. Sci Rep 2016;6:31448. https://doi.org/10.1038/srep31448.

[53] Bouslimani A, Porto C, Rath CM, Wang M, Guo Y, Gonzalez A, et al. Molecular cartography of the human skin surface in 3D. Proc Natl Acad Sci U S A 2015;112:2120–2129. https://doi.org/10.1073/pnas.1424409112.

[54] Ding T, Schloss PD. Dynamics and associations of microbial community types across the human body. Nature 2014;509:357–360. https://doi.org/10.1038/nature13178.

[55] Valm AM. The structure of dental plaque microbial communities in the transition from health to dental caries and periodontal disease. J Mol Biol 2019;431:2957–2969. https://doi.org/10.1016/j.jmb.2019.05.016.

[56] Rudney JD. Does variability in salivary protein concentrations influence oral microbial ecology and oral health? Crit Rev Oral Biol Med 1995;6:343–367. https://doi.org/10.1177/10454411950060040501.

[57] Dutzan N, Abusleme L, Bridgeman H, Greenwell-Wild T, Zangerle-Murray T, Fife ME, et al. On-going mechanical damage from mastication drives homeostatic Th17 cell responses at the oral barrier. Immunity 2017;46:133–147. https://doi.org/10.1016/j.immuni.2016.12.010.

[58] Dickson RP, Erb-Downward JR, Martinez FJ, Huffnagle GB. The microbiome and the respiratory tract. Annu Rev Physiol 2016;78:481–504. https://doi.org/10.1146/annurev-physiol-021115-105238.

[59] Wilbert SA, Mark Welch JL, Borisy GG. Spatial ecology of the human tongue dorsum microbiome. Cell Rep 2020;30:4003–4015.e3. https://doi.org/10.1016/j.celrep.2020.02.097.

[60] McLoughlin K, Schluter J, Rakoff-Nahoum S, Smith AL, Foster KR. Host selection of microbiota via differential adhesion. Cell Host Microbe 2016;1–10. https://doi.org/10.1016/j.chom.2016.02.021.

[61] Bunker JJ, Erickson SA, Flynn TM, Henry C, Koval JC, Meisel M, et al. Natural polyreactive IgA antibodies coat the intestinal microbiota. Science 2017;358. https://doi.org/10.1126/science.aan6619, eaan6619.

[62] Donaldson GP, Ladinsky MS, Yu KB, Sanders JG, Yoo BB, Chou W-C, et al. Gut microbiota utilize immunoglobulin A for mucosal colonization. Science 2018;360:795–800. https://doi.org/10.1126/science.aaq0926.

[63] Fadlallah J, El Kafsi H, Sterlin D, Juste C, Parizot C, Dorgham K, et al. Microbial ecology perturbation in human IgA deficiency. Sci Transl Med 2018;10. https://doi.org/10.1126/scitranslmed.aan1217, eaan1217.

[64] Weiser JN, Ferreira DM, Paton JC. Streptococcus pneumoniae: transmission, colonization and invasion. Nat Rev Microbiol 2018;1–13. https://doi.org/10.1038/s41579-018-0001-8.

[65] Moor K, Diard M, Sellin ME, Felmy B, Wotzka SY, Toska A, et al. High-avidity IgA protects the intestine by enchaining growing bacteria. Nature 2017;544:498–502. https://doi.org/10.1038/nature22058.

[66] Palm NW, de Zoete MR, Cullen TW, Barry NA, Stefanowski J, Hao L, et al. Immunoglobulin a coating identifies colitogenic bacteria in inflammatory bowel disease. Cell 2014;158:1000–10. https://doi.org/10.1016/j.cell.2014.08.006.

[67] Moll JM, Myers PN, Zhang C, Eriksen C, Wolf J, Appelberg KS, et al. Gut microbiota perturbation in IgA deficiency is influenced by IgA-autoantibody status. Gastroenterology 2021. https://doi.org/10.1053/j.gastro.2021.02.053.

[68] Jørgensen SF, Holm K, Macpherson ME, Storm-Larsen C, Kummen M, Fevang B, et al. Selective IgA deficiency in humans is associated with reduced gut microbial diversity. J Allergy Clin Immunol 2019;143:1969–1971.e11. https://doi.org/10.1016/j.jaci.2019.01.019.

[69] Donaldson GP, Lee SM, Mazmanian SK. Gut biogeography of the bacterial microbiota. Nat Rev Microbiol 2015;14:20–32. https://doi.org/10.1038/nrmicro3552.

[70] Briliūtė J, Urbanowicz PA, Luis AS, Baslé A, Paterson N, Rebello O, et al. Complex N-glycan breakdown by gut Bacteroides involves an extensive enzymatic apparatus encoded by multiple co-regulated genetic loci. Nat Microbiol 2019. https://doi.org/10.1038/s41564-019-0466-x.

[71] Cremer J, Arnoldini M, Hwa T. Effect of water flow and chemical environment on microbiota growth and composition in the human colon. Proc Natl Acad Sci U S A 2017;114:6438–43. https://doi.org/10.1073/pnas.1619598114.

[72] Randal Bollinger R, Barbas AS, Bush EL, Lin SS, Parker W. Biofilms in the large bowel suggest an apparent function of the human vermiform appendix. J Theor Biol 2007;249:826–31. https://doi.org/10.1016/j.jtbi.2007.08.032.

[73] Bergstrom K, Shan X, Casero D, Batushansky A, Lagishetty V, Jacobs JP, et al. Proximal colon–derived O-glycosylated mucus encapsulates and modulates the microbiota. Science 2020;370:467–72. https://doi.org/10.1126/science.aay7367.

[74] Donaldson GP, Chou W-C, Manson AL, Rogov P, Abeel T, Bochicchio J, et al. Spatially distinct

physiology of Bacteroides fragilis within the proximal colon of gnotobiotic mice. Nat Microbiol 2020. https://doi.org/10.1038/s41564-020-0683-3.

[75] Li H, Limenitakis JP, Fuhrer T, Geuking MB, Lawson MA, Wyss M, et al. The outer mucus layer hosts a distinct intestinal microbial niche. Nat Commun 2015;6:8292. https://doi.org/10.1038/ncomms9292.

[76] Sahakian AB, Jee SR, Pimentel M. Methane and the gastrointestinal tract. Dig Dis Sci 2010;55:2135–43. https://doi.org/10.1007/s10620-009-1012-0.

[77] Eswaran S, Muir J, Chey WD. Fiber and functional gastrointestinal disorders. Am J Gastroenterol 2013;108:718–27. https://doi.org/10.1038/ajg.2013.63.

[78] Zarrinpar A, Chaix A, Yooseph S, Panda S. Diet and feeding pattern affect the diurnal dynamics of the gut microbiome. Cell Metab 2014;20:1006–17. https://doi.org/10.1016/j.cmet.2014.11.008.

[79] Korem T, Zeevi D, Suez J, Weinberger A, Avnit-Sagi T, Pompan-Lotan M, et al. Growth dynamics of gut microbiota in health and disease inferred from single metagenomic samples. Science 2015;349:1101–6. https://doi.org/10.1126/science.aac4812.

[80] Gao Y, Li H. Quantifying and comparing bacterial growth dynamics in multiple metagenomic samples. Nat Methods 2018;15:1041–4. https://doi.org/10.1038/s41592-018-0182-0.

[81] Brown CT, Olm MR, Thomas BC, Banfield JF. Measurement of bacterial replication rates in microbial communities. Nat Biotechnol 2016;34:1256–63. https://doi.org/10.1038/nbt.3704.

[82] von Schwartzenberg RJ, Bisanz JE, Lyalina S, Spanogiannopoulos P, Ang QY, Cai J, et al. Caloric restriction disrupts the microbiota and colonization resistance. Nature 2021;1–6. https://doi.org/10.1038/s41586-021-03663-4.

[83] Stephen AM, Cummings JH. The microbial contribution to human faecal mass. J Med Microbiol 1980;13:45–56. https://doi.org/10.1099/00222615-13-1-45.

[84] Padmanabhan P, Grosse J, Asad Abu Bakar Md Ali, Radda GK, Golay X. Gastrointestinal transit measurements in mice with 99mTc-DTPA-labeled activated charcoal using NanoSPECT-CT. EJNMMI Res 2013;3(1):60. https://doi.org/10.1186/2191-219X-3-60.

[85] Costea PI, Hildebrand F, Arumugam M, Bäckhed F, Blaser MJ, Bushman FD, et al. Enterotypes in the landscape of gut microbial community composition. Nat Microbiol 2018;3:8–16. https://doi.org/10.1038/s41564-017-0072-8.

[86] Li J, Jia H, Cai X, Zhong H, Feng Q, Sunagawa S, et al. An integrated catalog of reference genes in the human gut microbiome. Nat Biotechnol 2014;32:834–41. https://doi.org/10.1038/nbt.2942.

[87] Ayeni FA, Biagi E, Rampelli S, Fiori J, Soverini M, Audu HJ, et al. Infant and adult gut microbiome and metabolome in rural Bassa and urban settlers from Nigeria. Cell Rep 2018;23:3056–67. https://doi.org/10.1016/j.celrep.2018.05.018.

[88] Schnorr SL, Candela M, Rampelli S, Centanni M, Consolandi C, Basaglia G, et al. Gut microbiome of the Hadza hunter-gatherers. Nat Commun 2014;5:3654. https://doi.org/10.1038/ncomms4654.

[89] Grizotte-Lake M, Zhong G, Duncan K, Kirkwood J, Iyer N, Smolenski I, et al. Commensals suppress intestinal epithelial cell retinoic acid synthesis to regulate interleukin-22 activity and prevent microbial dysbiosis. Immunity 2018;49:1103–1115.e6. https://doi.org/10.1016/j.immuni.2018.11.018.

[90] Atarashi K, Tanoue T, Oshima K, Suda W, Nagano Y, Nishikawa H, et al. Treg induction by a rationally selected mixture of Clostridia strains from the human microbiota. Nature 2013;500:232–6. https://doi.org/10.1038/nature12331.

[91] Moeller AH, Caro-Quintero A, Mjungu D, Georgiev AV, Lonsdorf EV, Muller MN, et al. Cospeciation of gut microbiota with hominids. Science 2016;353:380–2. https://doi.org/10.1126/science.aaf3951.

[92] Browne HP, Forster SC, Anonye BO, Kumar N, Neville BA, Stares MD, et al. Culturing of 'unculturable' human microbiota reveals novel taxa and extensive sporulation. Nature 2016;533:543–6. https://doi.org/10.1038/nature17645.

[93] Hildebrand F, Gossmann TI, Frioux C, Özkurt E, Myers PN, Ferretti P, et al. Dispersal strategies shape persistence and evolution of human gut bacteria. Cell Host Microbe 2021;29:1167–1176.e9. https://doi.org/10.1016/j.chom.2021.05.008.

[94] Caroll SB. The serengeti rules. Princeton University Press; 2016.

[95] Tett A, Pasolli E, Masetti G, Ercolini D, Segata N. Prevotella diversity, niches and interactions with the human host. Nat Rev Microbiol 2021. https://doi.org/10.1038/s41579-021-00559-y.

[96] Wibowo MC, Yang Z, Borry M, Hübner A, Huang KD, Tierney BT, et al. Reconstruction of ancient microbial genomes from the human gut. Nature 2021. https://doi.org/10.1038/s41586-021-03532-0.

[97] Tett A, Huang KD, Asnicar F, Fehlner-Peach H, Pasolli E, Karcher N, et al. The Prevotella copri complex comprises four distinct clades underrepresented in westernized populations. Cell Host Microbe 2019. https://doi.org/10.1016/j.chom.2019.08.018.

[98] Lynn DJ, Benson SC, Lynn MA, Pulendran B. Modulation of immune responses to vaccination by the microbiota: implications and potential mechanisms. Nat Rev Immunol 2021. https://doi.org/10.1038/s41577-021-00554-7.

[99] Xiao L, Feng Q, Liang S, Sonne SB, Xia Z, Qiu X, et al. A catalog of the mouse gut metagenome. Nat Biotechnol 2015;33:1103–8. https://doi.org/10.1038/nbt.3353.

[100] Gálvez EJCC, Iljazovic A, Amend L, Lesker TR, Renault T, Thiemann S, et al. Distinct polysaccharide utilization determines interspecies competition between intestinal Prevotella spp. Cell Host Microbe 2020;28:838–852.e6. https://doi.org/10.1016/j.chom.2020.09.012.

[101] Ramanan D, Bowcutt R, Lee SC, Tang MS, Kurtz ZD, Ding Y, et al. Helminth infection promotes colonization resistance via type 2 immunity. Science 2016;352:608–12. https://doi.org/10.1126/science.aaf3229.

[102] Chabé M, Lokmer A, Ségurel L. Gut protozoa: friends or foes of the human gut microbiota? Trends Parasitol 2017. https://doi.org/10.1016/j.pt.2017.08.005.

[103] Gabrielli S, Furzi F, Fontanelli Sulekova L, Taliani G, Mattiucci S. Occurrence of Blastocystis-subtypes in patients from Italy revealed association of ST3 with a healthy gut microbiota. Parasite Epidemiol Control 2020;9. https://doi.org/10.1016/j.parepi.2020.e00134, e00134.

[104] Li J, Fu R, Yang Y, Horz H-P, Guan Y, Lu Y, et al. A metagenomic approach to dissect the genetic composition of enterotypes in Han Chinese and two Muslim groups. Syst Appl Microbiol 2017. https://doi.org/10.1016/j.syapm.2017.09.006.

[105] Sonnenburg JL, Xu J, Leip DD, Chen C-H, Westover BP, Weatherford J, et al. Glycan foraging

in vivo by an intestine-adapted bacterial symbiont. Science 2005;307:1955–9. https://doi.org/10.1126/science.1109051.

[106] Cuskin F, Lowe EC, Temple MJ, Zhu Y, Cameron EA, Pudlo NA, et al. Human gut Bacteroidetes can utilize yeast mannan through a selfish mechanism. Nature 2015;517:165–9. https://doi.org/10.1038/nature13995.

[107] Ndeh D, Rogowski A, Cartmell A, Luis AS, Baslé A, Gray J, et al. Complex pectin metabolism by gut bacteria reveals novel catalytic functions. Nature 2017. https://doi.org/10.1038/nature21725.

[108] Rakoff-Nahoum S, Foster KR, Comstock LE. The evolution of cooperation within the gut microbiota. Nature 2016;533:255–9. https://doi.org/10.1038/nature17626.

[109] Mark Welch JL, Hasegawa Y, McNulty NP, Gordon JI, Borisy GG. Spatial organization of a model 15-member human gut microbiota established in gnotobiotic mice. Proc Natl Acad Sci U S A 2017;114:E9105–14. https://doi.org/10.1073/pnas.1711596114.

[110] Raman AS, Gehrig JL, Venkatesh S, Chang H-W, Hibberd MC, Subramanian S, et al. A sparse covarying unit that describes healthy and impaired human gut microbiota development. Science 2019;365. https://doi.org/10.1126/science.aau4735, eaau4735.

[111] Vangay P, Johnson AJ, Ward TL, Al-Ghalith GA, Shields-Cutler RR, Hillmann BM, et al. US immigration westernizes the human gut microbiome. Cell 2018;175:962–972.e10. https://doi.org/10.1016/j.cell.2018.10.029.

[112] Wexler AG, Goodman AL. An insider's perspective: Bacteroides as a window into the microbiome. Nat Microbiol 2017;2:17026. https://doi.org/10.1038/nmicrobiol.2017.26.

[113] Angelakis E, Bachar D, Yasir M, Musso D, Djossou F, Gaborit B, et al. Treponema species enrich the gut microbiota of traditional rural populations but are absent from urban individuals. New Microbes New Infect 2019;27:14–21. https://doi.org/10.1016/j.nmni.2018.10.009.

[114] Smits SA, Leach J, Sonnenburg ED, Gonzalez CG, Lichtman JS, Reid G, et al. Seasonal cycling in the gut microbiome of the Hadza hunter-gatherers of Tanzania. Science 2017;357:802–6. https://doi.org/10.1126/science.aan4834.

[115] Gomez A, Rothman JM, Petrzelkova K, Yeoman CJ, Vlckova K, Umaña JD, et al. Temporal variation selects for diet-microbe co-metabolic traits in the gut of Gorilla spp. ISME J 2016;10:514–526. https://doi.org/10.1038/ismej.2015.146.

[116] Tokuda G, Mikaelyan A, Fukui C, Matsuura Y, Watanabe H, Fujishima M, et al. Fiber-associated spirochetes are major agents of hemicellulose degradation in the hindgut of wood-feeding higher termites. Proc Natl Acad Sci U S A 2018;115:E11996–2004. https://doi.org/10.1073/pnas.1810550115.

[117] Bui AT, Williams BA, Hoedt EC, Morrison M, Mikkelsen D, Gidley MJ. High amylose wheat starch structures display unique fermentability characteristics, microbial community shifts and enzyme degradation profiles. Food Funct 2020;11:5635–5646. https://doi.org/10.1039/d0fo00198h.

[118] Li H, Gidley MJ, Dhital S. High-amylose starches to bridge the "Fiber gap": development, structure, and nutritional functionality. Compr Rev Food Sci Food Saf 2019;18:362–379. https://doi.org/10.1111/1541-4337.12416.

[119] Mizrahi I, Wallace RJ, Moraïs S. The rumen microbiome: balancing food security and environmental impacts. Nat Rev Microbiol 2021. https://doi.org/10.1038/s41579-021-00543-6, 0123456789.

[120] Xie H, Guo R, Zhong H, Feng Q, Lan Z, Qin B, et al. Shotgun metagenomics of 250 adult twins reveals genetic and environmental impacts on the gut microbiome. Cell Syst 2016;3:572–584. e3. https://doi.org/10.1016/j.cels.2016.10.004.

[121] Cwyk WM, Canale-Parola E. Treponema succinifaciens sp. nov., an anaerobic spirochete from the swine intestine. Arch Microbiol 1979;122:231–9. https://doi.org/10.1007/BF00411285.

[122] Adamovich Y, Ladeuix B, Sobel J, Manella G, Neufeld-Cohen A, Assadi MH, et al. Oxygen and carbon dioxide rhythms are circadian clock controlled and differentially directed by behavioral signals. Cell Metab 2019;29:1092–1103. https://doi.org/10.1016/j.cmet.2019.01.007.

[123] De Vadder F, Kovatcheva-Datchary P, Zitoun C, Duchampt A, Bäckhed F, Mithieux G. Microbiota-produced succinate improves glucose homeostasis via intestinal gluconeogenesis. Cell Metab 2016;24:151–157. https://doi.org/10.1016/j.cmet.2016.06.013.

[124] Koh A, De Vadder F, Kovatcheva-Datchary P, Bäckhed F. From dietary fiber to host physiology: short-chain fatty acids as key bacterial metabolites. Cell 2016;165:1332–1345. https://doi.org/10.1016/j.cell.2016.05.041.

[125] Kurilshikov A, Medina-Gomez C, Bacigalupe R, Radjabzadeh D, Wang J, Demirkan A, et al. Large-scale association analyses identify host factors influencing human gut microbiome composition. Nat Genet 2021;53:156–165. https://doi.org/10.1038/s41588-020-00763-1.

[126] Qin Y, Havulinna AS, Liu Y, Jousilahti P, Ritchie SC, Tokolyi A, et al. Combined effects of host genetics and diet on human gut microbiota and incident disease in a single population cohort. MedRxiv 2020. https://doi.org/10.1101/2020.09.12.20193045. 2020.09.12.20193045.

[127] Qi Q, Li J, Yu B, Moon J-Y, Chai JC, Merino J, et al. Host and gut microbial tryptophan metabolism and type 2 diabetes: an integrative analysis of host genetics, diet, gut microbiome and circulating metabolites in cohort studies. Gut 2021. https://doi.org/10.1136/gutjnl-2021-324053.

[128] Ségurel L, Bon C. On the evolution of lactase persistence in humans. Annu Rev Genomics Hum Genet 2017;18:297–319. https://doi.org/10.1146/annurev-genom-091416-035340.

[129] McKeen S, Young W, Fraser K, Roy NC, McNabb WC. Glycan utilisation and function in the microbiome of weaning infants. Microorganisms 2019;7. https://doi.org/10.3390/microorganisms7070190.

[130] Charlton S, Ramsøe A, Collins M, Craig OE, Fischer R, Alexander M, et al. New insights into Neolithic milk consumption through proteomic analysis of dental calculus. Archaeol Anthropol Sci 2019;11:6183–6196. https://doi.org/10.1007/s12520-019-00911-7.

[131] Day AJ, Cañada FJ, Díaz JC, Kroon PA, Mclauchlan R, Faulds CB, et al. Dietary flavonoid and isoflavone glycosides are hydrolysed by the lactase site of lactase phlorizin hydrolase. FEBS Lett 2000;468:166–170. https://doi.org/10.1016/s0014-5793(00)01211-4.

[132] Rühlemann MC, Hermes BM, Bang C, Doms S, Moitinho-Silva L, Thingholm LB, et al. Genome-wide association study in 8,956 German individuals identifies influence of ABO histo-blood groups on gut microbiome. Nat Genet 2021. https://doi.org/10.1038/s41588-020-00747-1.

[133] Ewald DR, Sumner SCJ. Blood type biochemistry and human disease. Wiley Interdiscip Rev Syst Biol Med 2016;8:517–535. https://doi.org/10.1002/wsbm.1355.

[134] Yamamoto F, Cid E, Yamamoto M, Blancher A. ABO research in the modern era of genomics. Transfus Med Rev 2012;26:103–118. https://doi.org/10.1016/j.tmrv.2011.08.002.

[135] Glass RI, Holmgren J, Haley CE, Khan MR, Svennerholm AM, Stoll BJ, et al. Predisposition for cholera of individuals with O blood group. Possible evolutionary significance. Am J Epidemiol 1985;121:791–6. https://doi.org/10.1093/oxfordjournals.aje.a114050.

[136] Aspholm-Hurtig M, Dailide G, Lahmann M, Kalia A, Ilver D, Roche N, et al. Functional adaptation of BabA, the H. pylori ABO blood group antigen binding adhesin. Science 2004;305:519–22. https://doi.org/10.1126/science.1098801.

[137] Moissl-Eichinger C, Pausan M, Taffner J, Berg G, Bang C, Schmitz RA. Archaea are interactive components of complex microbiomes. Trends Microbiol 2017. https://doi.org/10.1016/j.tim.2017.07.004.

[138] Ng F, Kittelmann S, Patchett ML, Attwood GT, Janssen PH, Rakonjac J, et al. An adhesin from hydrogen-utilizing rumen methanogen Methanobrevibacter ruminantium M1 binds a broad range of hydrogen-producing microorganisms. Environ Microbiol 2016;18:3010–21. https://doi.org/10.1111/1462-2920.13155.

[139] Brugère J-F, Borrel G, Gaci N, Tottey W, O'Toole PW, Malpuech-Brugère C. Archaebiotics: proposed therapeutic use of archaea to prevent trimethylaminuria and cardiovascular disease. Gut Microbes 2014;5:5–10. https://doi.org/10.4161/gmic.26749.

[140] Yeoman CJ, Thomas SM, Miller MEB, Ulanov AV, Torralba M, Lucas S, et al. A multi-omic systems-based approach reveals metabolic markers of bacterial vaginosis and insight into the disease. PLoS One 2013;8. https://doi.org/10.1371/journal.pone.0056111, e56111.

[141] Wang Z, Roberts AB, Buffa JA, Levison BS, Zhu W, Org E, et al. Non-lethal inhibition of gut microbial trimethylamine production for the treatment of atherosclerosis. Cell 2015;163:1585–95. https://doi.org/10.1016/j.cell.2015.11.055.

[142] Zhu W, Gregory JC, Org E, Buffa JA, Gupta N, Wang Z, et al. Gut microbial metabolite TMAO enhances platelet hyperreactivity and thrombosis risk. Cell 2016;165:111–24. https://doi.org/10.1016/j.cell.2016.02.011.

[143] Cho CE, Taesuwan S, Malysheva OV, Bender E, Tulchinsky NF, Yan J, et al. Trimethylamine-N-oxide (TMAO) response to animal source foods varies among healthy young men and is influenced by their gut microbiota composition: a randomized controlled trial. Mol Nutr Food Res 2017;61:1–12. https://doi.org/10.1002/mnfr.201600324.

[144] Chhibber-Goel J, Gaur A, Singhal V, Parakh N, Bhargava B, Sharma A. The complex metabolism of trimethylamine in humans: endogenous and exogenous sources. Expert Rev Mol Med 2016;18. https://doi.org/10.1017/erm.2016.6, e8.

[145] Mao J, Zhao P, Wang Q, Chen A, Li X, Li X, et al. Repeated 3,3-dimethyl-1-butanol exposure alters social dominance in adult mice. Neurosci Lett 2021;758:136006. https://doi.org/10.1016/j.neulet.2021.136006.

[146] Li Q, Korzan WJ, Ferrero DM, Chang RB, Roy DS, Buchi M, et al. Synchronous evolution of an odor biosynthesis pathway and behavioral response. Curr Biol 2013;23:11–20. https://doi.

org/10.1016/j.cub.2012.10.047.

[147] Apps PJ, Weldon PJ, Kramer M. Chemical signals in terrestrial vertebrates: search for design features. Nat Prod Rep 2015;32:1131–53. https://doi.org/10.1039/c5np00029g.

[148] Mukherji A, Kobiita A, Ye T, Chambon P. Homeostasis in intestinal epithelium is orchestrated by the circadian clock and microbiota cues transduced by TLRs. Cell 2013;153:812–27. https://doi.org/10.1016/j.cell.2013.04.020.

[149] Reitmeier S, Kiessling S, Clavel T, List M, Almeida EL, Ghosh TS, et al. Arrhythmic gut microbiome signatures predict risk of type 2 diabetes. Cell Host Microbe 2020;28:258–272.e6. https://doi.org/10.1016/j.chom.2020.06.004.

[150] Thaiss CA, Levy M, Korem T, Dohnalová L, Shapiro H, Jaitin DA, et al. Microbiota diurnal rhythmicity programs host transcriptome oscillations. Cell 2016;167:1495–1510.e12. https://doi.org/10.1016/j.cell.2016.11.003.

[151] Leone V, Gibbons SM, Martinez K, Hutchison AL, Huang EY, Cham CM, et al. Effects of diurnal variation of gut microbes and high-fat feeding on host circadian clock function and metabolism. Cell Host Microbe 2015;17:681–9. https://doi.org/10.1016/j.chom.2015.03.006.

[152] Yoon HS, Cho CH, Yun MS, Jang SJ, You HJ, Hyeong KJ, et al. Akkermansia muciniphila secretes a glucagon-like peptide-1-inducing protein that improves glucose homeostasis and ameliorates metabolic disease in mice. Nat Microbiol 2021;1–11. https://doi.org/10.1038/s41564-021-00880-5.

[153] Chevalier C, Stojanović O, Colin DJ, Suarez-Zamorano N, Tarallo V, Veyrat-Durebex C, et al. Gut microbiota orchestrates energy homeostasis during cold. Cell 2015;163:1360–74. https://doi.org/10.1016/j.cell.2015.11.004.

[154] Ouwerkerk JP, van der Ark KCH, Davids M, Claassens NJ, Finestra TR, de Vos WM, et al. Adaptation of Akkermansia muciniphila to the oxic-anoxic interface of the mucus layer. Appl Environ Microbiol 2016;82:6983–93. https://doi.org/10.1128/AEM.01641-16.

[155] Adamovich Y, Ladeuix B, Golik M, Koeners MP, Asher G. Rhythmic oxygen levels reset circadian clocks through HIF1α. Cell Metab 2017;25:93–101. https://doi.org/10.1016/j.cmet.2016.09.014.

[156] Li B, He Y, Ma J, Huang P, Du J, Cao L, et al. Mild cognitive impairment has similar alterations as Alzheimer's disease in gut microbiota. Alzheimers Dement 2019;1–10. https://doi.org/10.1016/j.jalz.2019.07.002.

[157] Wu H, Esteve E, Tremaroli V, Khan MT, Caesar R, Manneras-Holm L, et al. Metformin alters the gut microbiome of individuals with treatment-naive type 2 diabetes, contributing to the therapeutic effects of the drug. Nat Med 2017;23:850–8. https://doi.org/10.1038/nm.4345.

[158] Jie Z, Xia H, Zhong SL, Feng Q, Li S, Liang S, et al. The gut microbiome in atherosclerotic cardiovascular disease. Nat Commun 2017;8:845. https://doi.org/10.1038/s41467-017-00900-1.

[159] Thaiss CA, Zeevi D, Levy M, Zilberman-Schapira G, Suez J, Tengeler AC, et al. Transkingdom control of microbiota diurnal oscillations promotes metabolic homeostasis. Cell 2014;159:514–29. https://doi.org/10.1016/j.cell.2014.09.048.

[160] Kuang Z, Wang Y, Li Y, Ye C, Ruhn KA, Behrendt CL, et al. The intestinal microbiota programs diurnal rhythms in host metabolism through histone deacetylase 3. Science 2019;365:1428–34.

https://doi.org/10.1126/science.aaw3134.

[161] Zwighaft Z, Aviram R, Shalev M, Rousso-Noori L, Kraut-Cohen J, Golik M, et al. Circadian clock control by polyamine levels through a mechanism that declines with age. Cell Metab 2015;22:874–85. https://doi.org/10.1016/j.cmet.2015.09.011.

[162] Johnson CH, Zhao C, Xu Y, Mori T. Timing the day: what makes bacterial clocks tick? Nat Rev Microbiol 2017;15:232–242. https://doi.org/10.1038/nrmicro.2016.196.

[163] Paulose JK, Wright JM, Patel AG, Cassone VM. Human gut bacteria are sensitive to melatonin and express endogenous circadian rhythmicity. PLoS One 2016;11. https://doi.org/10.1371/journal.pone.0146643, e0146643.

[164] Paulose JK, Cassone CV, Graniczkowska KB, Cassone VM. Entrainment of the circadian clock of the enteric bacterium Klebsiella aerogenes by temperature cycles. IScience 2019;19:1202–1213. https://doi.org/10.1016/j.isci.2019.09.007.

[165] Eelderink-Chen Z, Bosman J, Sartor F, Dodd AN, Kovács ÁT, Merrow M. A circadian clock in a nonphotosynthetic prokaryote. Sci Adv 2021;7. https://doi.org/10.1126/sciadv.abe2086, eabe2086

第 3 章

宏基因组样本收集

　　摘　要：高质量的样本是进行宏基因组学研究的先决条件。早些年，辛勤的学生每天早上都要在医院里等待，以确保采集到的每一勺粪便样本都能保存在冰冻的无菌管中，以最快速度冷冻于 –80℃的冰箱，随后装上大量干冰分批运输。本章涉及对样本采集步骤的建议，从样本收集、测序、统计分析时需要考虑的因素。除了宏基因组测序外，还需要多种方法来评估低生物量样本中的微生物数量，并将微生物与潜在功能联系起来。污染控制和身体多采样位置的研究设计的原则同样适用于各种微生物组样本。

　　关键词：宏基因组鸟枪法测序，扩增子测序，核糖体 rRNA 基因，微生物组，低生物量标本，胎盘微生物组，脂肪组织，宏基因组全关联研究，标本储存

3.1　样本中非微生物的部分，会影响DNA提取和测序量

　　根据布里斯托（Bristol's stool score，BSS）大便分类法，可根据形状对粪便样本打分，这是对样本中水分含量、肠道经过时间以及样本中微生物数量的有效估计[1-3]。被试者可根据问卷备注给出粪便样本 BSS 打分。问卷得出的结果和粪便宏基因组样本中得到的次级胆汁酸代谢酶数量相关，而这在机制上是说得通的[3]。

　　自动采样马桶可以自动记录诸如 BSS 得分、尿样的体积和流速，但还不能自动采集样本[4]。

　　每克宏基因组样本中的微生物数量，受到食物残渣的影响，这对于诸如大熊猫这类食草动物尤其明显（图 3.1A）。一项基于英国健康成人饮食的粪便研究，在剧烈震荡和洗涤剂处理后，发现样本中 55% 为细菌、17% 为纤维、24% 为可溶性物质。随着微生物群落的不同，样本中细菌所占的比例会有所差异。

　　可折断的尼龙拭子，能被用于粪便和口腔样本采样（图 3.1B）。皮肤或鼻腔采样常使用拭子，采样前通常会用生理盐水或缓冲液浸润，但目前尚不清楚短暂接触矿物质会对微生物群落有多少影响。对于生物量较低的样本，需要的采样量会更高。为避免样本在运输过程中漏液，需谨记拧紧瓶盖。如果样本是受试者自采集的，尽量配有清晰的图片及视频说明采样流程，并对样本拍照。不同队列间受试者的配合程度不

同，需要尽早开展质控环节。

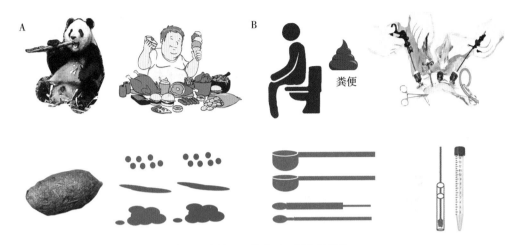

图 3.1　在采集宏基因组样本时，需获得足量 DNA

A. 宏基因组样本可能含有除微生物之外的其他物质。人类粪便按照简化的BSS打分描述，最硬的BSS得分为1，中间硬度的为4，水分含量最多的为7。B. 采集样本时，固体样品可用勺子，表面可用拭子，液体可用管子。腹腔镜或者其他新技术可以在手术室采集微生物组样本，降低切口污染的概率。皮肤和其他表面采样前都要经过净化。棉签或刷子可以在保护管之前到取样地点。

若受试者患有痔疮或炎症性肠炎等胃肠道疾病，粪便样本中会包含血液，导致宏基因组样本中人类基因组的比例远超 1%（表 3.1）。除了粪便和龈上菌斑，大部分人类宏基因组样本中都包含更高比例的来自人类基因组的序列，甚至在某些组织的样本中高达99%。在签署知情同意书后，宏基因组获得的人类低深度全基因组测序数据，可以与微生物数据共同分析，但需要注意不同来源组织与来自血液的人基因组数据存在差异。

表 3.1　来自不同采样部位的人类宏基因组样本中，采用短序列测序时，包含的人源 DNA 序列占比

采样区域	采样点	人源基因比例	参考文献
肠道	粪便	1%	[5,6]
肠道	粪便（克罗恩病）	20% 或以上	[7]
口腔	颊黏膜	82%～90%	[5,8,9]
口腔	龈上菌斑	40%，5.6%	[5,10]
口腔	龈下菌斑	79%	[5]
口腔	舌苔	30%	[9,11]
口腔	唾液	77%～91%	[5,11]
皮肤	干燥（如前臂）	36%	[12]
皮肤	潮湿（如肘窝）	44%	[12]
皮肤	皮脂（如耳后皱褶）	59%～73%	[5,12]
泌尿生殖道	阴道	90%～98%	[5,13]
泌尿生殖道	子宫颈口	98%	[14]
泌尿生殖道	腹水	99.8%	[15]

续表

采样区域	采样点	人源基因比例	参考文献
泌尿生殖道	胎盘	>99%	[16]
呼吸道	前鼻孔	96%	[5]

注：这并不是一个详尽的列表。当研究人员和临床医生决定宏基因组NGS的测序数量（例如，10 GB的PE100的读数），或选择扩增子测序、原位杂交等方法时，该表可提供参考。

通过低速离心去除宿主细胞的方式会导致某些细菌物种的消失，这在支气管肺泡灌洗（bronchoalveolar lavage）样本中已被论证[17]。通过分子生物学或者化学手段去除宿主细胞，也会影响微生物的组成[15-18]，但这可在未来进行优化。目前我们推荐进行无偏测序，之后使用生物信息学手段去除宿主。

根据研究问题选择适当的技术组合（如胰腺癌中的细菌，图 1-9）。对于采集完备的样本，若仅仅因为数据量过大而不去穷尽其分析潜力则很可惜。毕竟，很多微生物的演化和互作发生在局部。

3.2 对于粪便及生物量较低的宏基因组，要注意减少采样过程中每一步的污染

正如在第 1 章简要提及的，对于生物量较低的宏基因组，要注意采样过程中每一步的污染风险，例如对胎盘样本是否包含微生物存疑（图 1.8），但对于肿瘤样本接受度就更高（图 1.9）。研究生物量较低的宏基因组样本，有助于建立在采样过程中每一步注意事项的清晰认识（图 3.2）。另外，证明某一部位缺少微生物，比证明微生物的存在更难。毕竟大自然中无论怎样极端的环境，细菌、真菌和古菌总能占有一席之地。

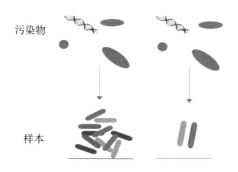

图 3.2 采样过程中需注意事项

从采样到测序的任何步骤都可能引入污染的DNA或微生物。只是生物量低的样品更容易被污染物淹没

对于粪便、口腔和阴道样本，通常拭子中包含超过 10^{10} CFU 微生物细胞。庞大的细胞数使我们在样本采集、DNA 提取、文库制备及测序过程中，无需为避免污染

而过度谨慎担忧。而对于生物量较低的样本（图3.2和图3.3），需要经常检验上述步骤所用试剂是否包含活的或死的微生物[25]。例如，使用扩增子测序（图3.4）。关于样本污染，经常被引的一篇文献使用连续稀释的沙门氏菌，在五次连续稀释之后，即包含约1000个沙门菌细胞时，16S测得的其他菌看上去更多了[26]。尽管实验关注的是试剂污染，作者并没有报道有关稀释过程是否引入了污染的信息（PCR扩增是在一个超净工作台里用高压灭菌微量离心管和滤芯移液吸头进行的），也没有报道有采用什么措施避免样本被操作人员污染，这在古DNA领域中众所周知。在进行扩增子测序时，还需要考虑批次效应带来的影响[27]。

哪怕是医院的房间，甚至包括空间站，在使用后也包含微生物，它们可能来自患者和员工[28-32]。医院的通风系统中，可能检出细菌或真菌（图3.3），当其与我们关注的样本存在差异时，仍可以认为样本中包含微生物，并寻找进一步的证据。对于作为阴性对照的培养皿上的活菌群以及qPCR，上述结论也适用。

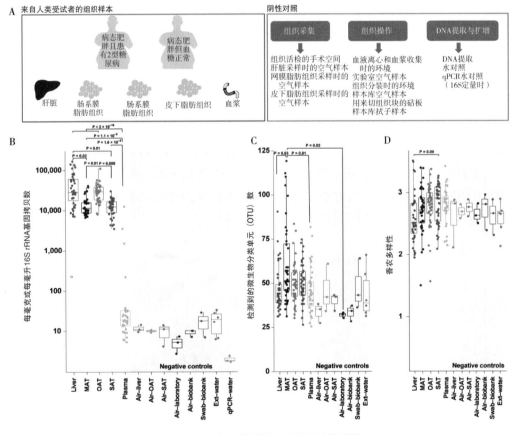

图3.3　一个低生物量样本研究设计的例子

包括阴性对照以及不同的生理条件下的关联研究。A. 采集病态肥胖者，2型糖尿病T2D（$n=20$）和正常血糖（$n=20$）的肝脏、3个不同的脂肪组织（OAT、MAT、SAT）和血浆样本。采用优化的血浆和组织细菌DNA检测条件进行DNA提取和扩增。在分析的主要步骤：组织收集、组织操作、DNA提取和扩增，采用一套全面的阴性质控样本，来检测环境带来的样品污染。在组织收集过程中，输卵管在整个手术过程中（空气-肝脏、空气-AOAT和空气-SAT）一直开放在手术区附近。来自组织操作的污染由另一组管子予以避免，在整个血液离心和等离子体收集（空气实验室）以及组织外引用（空气生物库）过程中，这些管子一直开放在操作者旁边。用于检验组织的砧板在组织操作（拭子生物库）之前被取样。采用定量PCR技术，使用纯水样本对DNA提取（ext-water）过程中的标本、试剂和（或）环境污染进行质控，并对组织16S RNA进行扩增定量分析。在逐个病例对阴性对照进行全面验证后，16S定量和测序数据被用于发现与T2D相关的组织特异性细菌特征。（B～D）身体各部位的细菌数量。B. 16S rRNA基因计数。C. 观察到的OTU分类。D. 肝脏中的香农指数、三种不同的脂肪组织（OAT、MAT和SAT）和肥胖者的血浆。在分析的主要步骤：组织采集（空气-肝脏、空气-OAT、空气-SAT）、组织操作（空气-实验室、空气-生物库和棉签生物库）和DNA提取或扩增（纯水、qPCR-水）中，检测阴性对照以控制环境样本污染。在图（B～D）中，各组用Kruskal-Wallis单因素方差分析进行比较，然后用Dunn的成对检验进行比较。方框图描绘了第一和第三个四分位数，中位数由方框内的垂直线表示；线段分别从第一和第三个四分位数延伸到最高和最低观测值，不超过$1.5 \times IQR$。来源：Nat Metab, 2020, 2:233-242. https://doi.org/10.1038/s42255-020-0178-9.

对于包含大量人源DNA的样本，扩增过程可能会非特异性地扩增大量人源序列（如肾结石样本）[33]，除了最优化PCR条件，这样的样本还需要根据DNA片段长度进行纯化，或在控制污染的前提下，进行目标区域测序。

对于来自同一人的多份样本，包括采集自不同部位的样本，也能够识别个人特有的模式，从而有助于排除采集到的微生物是随机污染，或是来自于试剂的系统性污染[15,34,35]。当研究没有怀孕也没有炎症的女性生殖道样本时，笔者合作的临床专家吴瑞芳教授确保在皮肤上进行亚厘米的切割后，她的团队从道格拉斯窝中的腹腔液（盆腔积液）开始，将腹腔镜转移到输卵管（如输卵管有梗阻的患者），然后是子宫内膜（子宫肌瘤、子宫内膜异位症或子宫腺肌病），根据生物量由低到高，逐个进行采样，阴道和宫颈样本在首次就诊当天采集。在操作时有保护性套管，到达采样部位时才伸出采样。尽管乳酸杆菌占优势，但是阴道和宫颈样本与几天后在手术室为同一位志愿者收集的上生殖道样本的微生物分布有很好的一致性。腹腔液pH相比更加接近中性，也可能是低生物量但多样性更高微生物的来源（第2章）[34,35]。

目前，胎盘中是否存在微生物，仍然存在争议（图1.8）。根据对保守的16S RNA区域进行原位杂交的结果，Kjersti M. Aagaard博士报告说，小团细菌主要位于绒毛薄壁组织或合胞体滋胚层，而在绒毛膜和母体绒毛间隙中较少见[36]。图3.4展示了核糖体RNA（rRNA）基因簇（rDNA）的示意图。最近的一项研究从胎盘绒毛末端采集标本，要求能使用16S rRNA基因扩增子测序（V1～V2区，图3.4）和宏基因组短序列测序都能检测到细菌的存在，该研究否认了在同一产妇的阴道样本中也包含的细菌[25]。这样各种删结果之后，只有新生儿病原性无乳链球菌（B型链球菌），被认定为胎盘微生物[14]。

来自血液的污染一直是一个难以回避的问题（图3.3）。在胎盘研究中，由于测序量极低［平均2650万条读段（reads），其中超过99%是人源的，如果99.9%都来自人源，那么每个样本平均只剩下2.65万条读段，相当于每个细菌的基因组覆盖度不足 1%〕，导致从 KEGG 第二层看到的功能分布，存在样本间的波动[16]。对于宫颈口样本，我们发现优势菌种与人源序列百分比呈现相关性[9,14]。然而，Aagaard 博士的首篇研究中确定的一些分类单元，可能仍然真正地存在于胎盘中。这项宏基因组学研究中，胎盘微生物组显示出一种丰度较高的大肠埃希菌存在，该菌已知在新生儿粪便中出现[37-41]，同时还包含一些可能来自肠道或口腔的细菌，如拟杆菌属、副血链球菌、普雷沃黑色素细菌，以及一些可能来自生殖道的细菌，如痤疮丙酸杆菌、惰性乳杆菌、卷曲乳杆菌等，这些细菌都在包含众多控制样本的实验组中出现。值得注意的是，痤疮丙酸杆菌和副血链球菌也是婴儿口腔和肠道微生物组的成员[16,38,40,42]。如果对奈瑟菌的物种分配是正确的，尽管测序覆盖率低[16]（更多关于分类学的内容见第 5 章），它将为幼儿鼻咽部携带的细菌提供一个潜在的来源。16S rRNA 扩增子测序[36]检测到了更多的分类群，这样的测序方式不会包含人类序列。

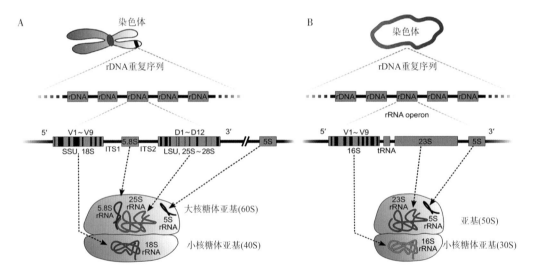

图 3.4　核糖体 RNA（rRNA）基因簇（或 rDNA）的示意图

真核生物和原核生物RNA位点的可变区通常用于表征微生物的分类群，并通过扩增子测序和分析来找出它们之间的系统发育关系。在大多数真菌中，rRNA基因簇包括小核糖体亚基（SSU，18S）和大核糖体亚基（LSU，25~28S），其中内转录间隔区（ITS1和ITS2）位于5.8S的侧翼。在细菌中，rRNA操纵子包括SSU（16S）、LSU（23S）和5S位点。黑色的垂直线以串行顺序显示了SSU（V1~V9）和LSU（D1~D12）中的可变区域，最适合通过微生物群落剖面进行生物多样性评估。来源：Trends Microbiol, 2021, 29:19-27. https://doi.org/10.1016/j.tim.2020.05.019.

根据对 T2D 患者脂肪组织细菌及其异常菌群的有趣研究，血浆和阴性对照组细菌多样性并不低于脂肪组织样本，但 16S rRNA 基因总拷贝数高于阴性对照组（图 3.3）[43,44]。每毫克组织[44]中含有的细菌数量，看起来比每微克（μm）单位 DNA[43]的细菌数量更多，因为后者包括人类基因组（3.2 GB 的 2 个拷贝）。白色脂肪细胞的直径从小于 30 μm 到大于 300 μm 不等[45]。回到第一章关于人体细胞与细菌比例的问题，我们可以尝试一些估算。如果这些都是直径为 20 μm 的微小脂肪细胞，那么根据图 3.3 中 16S RNA 基因拷贝数[44]，单位组织重量或体积中，细菌和人体细胞的比率大约为 1∶10。然而，如果脂肪细胞直径为中位数，约为 100 μm，细菌细胞与人类细胞的比例就会反过来，成为 10∶1 左右。

当开始研究大脑和肺部（图 3.5、图 3.6）或其他组织时，我们是否按照相同的标准，收集了所有相关的样本？好消息是对于支气管肺泡灌洗液，在有保护性插管时，从口腔和从鼻腔采集的样本之间，没有明显的差异[46]。但是，相关的样本应该在个体间进行两两比较，而不是显示在一个粗糙的 PCA（主成分分析）中。这样的样本中会有多少微生物细胞？对于病毒和真菌，我们是否需要其他信息来更好地了解其当地栖息地和形态学特征？

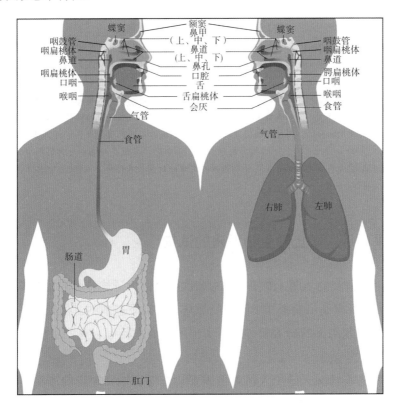

图 3.5　微生物在人体内的分布情况

微生物可能从鼻子、口腔或咽喉扩散到远端身体部位的示意图，以及不同部位之间的相互联系。例如，咽扁桃体（又称腺样体），是口腔/鼻咽部淋巴组织的一个主要部位，可能成为中耳感染的感染菌的储备来源，是因为它们可以通过咽鼓管传播。来源：Cell Host Microbe, 2017, 21:421-432. https://doi.org/10.1016/j.chom.2017.03.011.

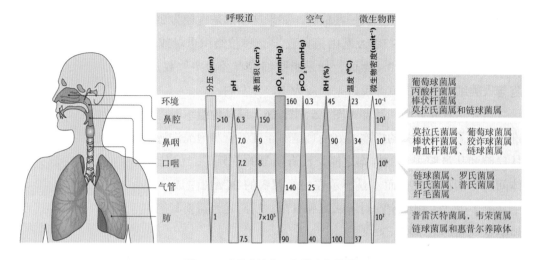

图 3.6 呼吸道的生理和微生物梯度

呼吸道从鼻腔、鼻咽、口咽、气管到肺部,具有不同的生理和微生物的梯度特征。pH 随呼吸道逐渐升高,而相对湿度和温度的升高主要发生在鼻腔。此外,氧气和二氧化碳的分压呈相反的梯度,这取决于环境空气条件和肺表面的气体交换。呼吸道还受到来自环境的颗粒的影响,其中包括细菌和病毒。颗粒的沉积位置与其直径有关,大于 10 μm 的颗粒主要沉积在呼吸道上部,小于 1 μm 的颗粒可以进入肺部。这些颗粒通常含有细菌和病毒,其直径一般大于 0.4 μm。呼吸道的生理参数决定了微生物在不同生态位的选择性生长的条件,从而形成了呼吸道不同位置的微生物群落。测量细菌密度的单位根据生态位不同而有所差异,可用以下方式表示:环境中的细菌密度可以用每立方厘米空气中的细菌数来表示,而鼻腔、鼻咽、口咽和肺部的细菌密度则可以用每个鼻拭子、每毫升口腔清洗液或每毫升支气管肺泡灌洗液中的细菌数来估计。来源:Nat Rev Microbiol, 2017, 15:259-70. https://doi.org/10.1038/nrmicro2017.14。

3.3 采样后,防止微生物增殖的试剂

在宏基因组研究的早期,粪便样本中兼性厌氧性大肠埃希菌的相对丰度有时超过 30%,研究人员怀疑是由粪便样本长时间暴露在室温下造成的。但是这一现象也可能反映了一些疾病状态下的肠道菌群的真实情况,如大肠癌、克罗恩病、IgA 缺乏、2 型糖尿病等[7,47-50]。

为了避免这类情况,研究人员不再需要让志愿者们在家里的冰箱里暂存粪便,或者每天在诊所里放干冰。冷冻过程也会影响宏基因组样品的组成,例如,由于水结晶过程中 pH 和其他浓度的变化,宏基因组样品中的组分可能会影响冷冻效率。目前,有一些商业试剂可使微生物组样品在室温下保存 2 ~ 4 周(如 DNA Genotek Inc., Mawi DNA Technologies, MGI Tech 提供的产品)。这些试剂的保存时间通常远远超过快递公司的运送时间。过滤纸也可用于粪便和宫颈样本,取样后风干,然后密封,但在滤纸风干、裁剪等过程中如何最大限度地减少污染,并没有形成共识。我们必须确保滤纸上的 DNA 数量足够进行高通量宏基因组测序,而不仅仅是使用 16S rRNA

基因扩增子测序。

对于宏转录组的研究，目前还没有发表足够的文献，这不仅需要高质量的保存RNA，还要求后续（无论多么不完全）去除核糖体RNA[51,52]。宏蛋白质组学也在兴起，我们目前只尝试了新鲜或冷冻的样品。

稳定剂只抑制细菌生长和降解，不杀灭细菌。因此，仍有一些微生物可以在培养皿上生长。另外，使用质谱法分析相同样本的代谢组，通常需要将（尼龙）拭子浸泡在与存储微生物组不同的试剂中（如用于皮肤拭子代谢组检测的乙醇与水50∶50混合液[53]），但是也有一些商业产品试图兼顾这两种用途。

思考题 3.1

（1）对于一个你感兴趣的微生物样本，你知道它在身体的哪个部位，以及那里大概有多少微生物细胞和物种吗？当患上某种疾病时，你认为这些数字将如何改变？

（2）如果你想构建一个宏基因组文库，你需要准备多少DNA（例如0.5 μg）？你会使用拭子还是塑料采样器来采集样本？你能否立刻处理或冷冻样本（之后用于其他组学分析）？如果你使用一个商业试剂，在室温下保存样本数周，并通过商业快递运输，你认为这会不会对微生物群落中的某些菌群产生较大的影响？

（3）你是否有一个标准的混合后的真实样本，用来与真实的、冷冻保存的和保存液保存的样本之间的微生物群落进行对比？

3.4 对于宏基因组样本的DNA提取方法

宏基因组样本的DNA提取比哺乳动物细胞或单一微生物种类的DNA提取更具挑战性，因为它涉及破坏多种不同类型的细胞壁（图3.7）[54-56]，即使样本中没有过多的植物纤维或原生生物（图3.1）。为了提高DNA提取的效率和质量，可以采用物理和化学的方法，同时使用对照样品进行平行测序。还可以掺入与样本中序列无关的质控品用于定量。如果需要进行纳米孔测序，要求片段长度超过20kb，那么就需要采用更加温和的提取方法。

珠磨法是一种常用的手段，它使用坚硬的四方锆多晶体珠子来破碎细胞壁，比玻璃珠更有效。珠子的大小也要根据样品中的细菌和真菌的特点进行选择。较小的珠子可以在单位时间内产生更多的碰撞，但可能无法完全破坏一些真菌的细胞壁。

与扩增子测序不同，高通量宏基因组测序不涉及PCR步骤，因此对DNA的纯度要求不那么严格。例如，从粪便样本中提取的DNA可能仍然会有点发黄，但这不会影响测序的结果。

此外，还可以使用自动化平台（如96孔板）来提高样品处理的质量和一致性，减少随机污染的风险,并节省人力和时间。这种自动化还可以更好地保护工作人员免受任何临床样本污染。

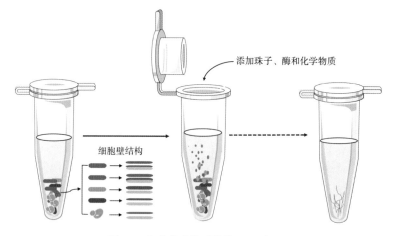

图 3.7　复杂微生物群落的 DNA 提取

粗线条代表革兰氏阴性菌、革兰氏阳性菌和真菌的不同细胞壁结构。

对于构建测序文库,提取的 DNA 片段要经过超声打断。使用诸如 Tn5 酶进行酶纯化,通量更高,也更适合自动化。

3.5　测序量

理论上,只要一个测序读取能够唯一地比对到一个分类单元,就可以检测出特定微生物的存在(详见第 5 章关于分类的内容)的存在。宏基因组鸟枪法测序不需要 PCR 扩增步骤,因此其错误率可以忽略不计。对于一个包含 1 亿个读数的宏基因组样本(如读长 PE100,即 $100 \times 10^6 \times 100 = 10$ GB 数据),直接检测到的一个分类群或基因的最低可能相对丰度是 10^{-8}。对于含 10^{11} CFU 微生物细胞的样品,检出限应该达到单细胞,可以用连续稀释的对照[22,60]进行验证。对于低丰度类群,测序量仍然是一个限制因素。在这种情况下,可以考虑在同一个体其他部位样本或其他个体的样本中寻找目标微生物,或者首先尝试对样本进行培养。

DNA 作为一种长链大分子,遵循长链一端固定时或弯或直形态分布的物理规律,桥式 PCR 测序平台倾向于对高 GC 区域(如双歧杆菌基因组的 GC 含量约为 60%)进行过度测序,并且可能需要对过度测序导致的丰度偏差进行数值校正[57,61,62]。

对于细菌的 16S RNA 扩增子测序,以及真菌的 ITS(内转录间隔区)基因扩增子测序(图 3.4),即使在 DNA 浓度低于紫外线检测(例如,在一些尿样中)的下限时,仍可以检测到微生物。不同的高变区(如 V4 ~ V5,V1 ~ V2)具有不同的分类学分辨率。全长扩增子测序,即包含了完整 16S 和 18S ITS 序列的测序,是一种可靠的微生物分类方法[63]。

思考题 3.2

样本	DNA 含量（ng/μL）	体积（μL）	总 DNA 量（μg）
A1	8.56	80	0.68
A2	1.32	80	0.11
A3	24.4	80	1.95

（1）上表是来自三个样本的 DNA 提取结果。如果粪便样本中，平均基因组大小为 5Mb，每个样本中大约有多少细菌？

（2）如果这是每个细菌基因组平均 2.3 Mb，平均有 96% 的人源 DNA 的阴道样本，每个样本中预期会有多少细菌细胞？

（3）如果这些阴道样本中，主要的菌是惰性乳酸杆菌，其基因组大小 1.3 Mb，每个样本中会包含多少个细菌细胞呢？

（4）如果这些阴道样本主要包含白色念珠菌这一种真菌，在每个样本中估计包含多少个真菌基因组呢？

3.6 分类和功能概况，以及绝对丰度

从测序数据中，我们可以通过基因（详见第 5 章）得出分类和功能概况。在分类任务中，需要注意的是，某些研究较少的人类反转录序列，可能会被误认为是 RNA 病毒。

当将相对丰度加和后正则化为 1 时，请注意未分类的部分，该比例在某些样本中是偏高的。

经典的功能注释数据库是 KEGG[64]，其中包含了一些新增的模块，例如脑肠轴功能模块[65,66]。很多 KO（KEGG 同源群）在多个功能模块中都存在。当我们报告某一样本具有特定模块时，通常采用 60% 这一阈值（检测到了模块中 > 60% 以上的 KO，即认为该模块真实存在）。另外，KO 只是粗略比对，一个酶真正的功能，可能改变一个关键氨基酸就会不一样，例如代谢产生丙酸咪唑的酶[67]。在这领域，还有很多基础研究有待补足。

如果在以上每一步都做好了控制，可以估计一个样本中微生物的绝对数量。最近一项关于早产儿粪便微生物组的研究使用了掺入质控细胞（spike-in cell），并报告了在绝对丰度的动态变化。例如，真菌白色念珠菌似乎抑制了许多细菌，或者更确切地说，当它没有被细菌有效地竞争时，它就开始繁殖。所有样品中本来都不含有用于掺入的细菌（红色沙林杆菌 DSM 13855），古菌（西班牙海囊藻 ATCC 33960），以及

真菌（瑞氏木霉 Trichoderma reesei ATCC 13631）。这些物种能够被检出，是因为它们定量地被加入粪便样本中。

思考题 3.3

在下面的研究中找到对应的功能概述（使用 16S 扩增子测序），并尝试绘出母乳样本（有些样本使用了奶泵）与作为阴性对照样本之间的差异。除了这个表，你认为还有什么是合适的可视化工具？基于 Bray-Curt 的 PCoA，或是每个分类单元的箱线图或小提琴图？

在 CHiLD（加拿大健康婴儿纵向发展）队列中 393 位母亲之间的"核心乳汁微生物群"与阴性对照的比较（表 S3 见 [69]）

	样本（$n=393$）				阴性对照
谱系	属	均值±标准差	最大值	流行率	流行率
Proteobacteria—Burkholderiales	*Unclassified*	5.86±3.43	12.56	100%	13%
Firmicutes—Staphylococcaceae	*Staphylococcus*	4.86±11.5	87.5	100%	20%
Proteobacteria—Oxalobacteraceae	*Ralstonia*	4.79±2.76	9.41	100%	7%
Proteobacteria—Comamonadaceae	*Unclassified*	4.42±2.58	9.75	100%	7%
Proteobacteria—Comamonadaceae	*Acidovorax*	3.95±2.34	13.33	100%	20%
Proteobacteria—Oxalobacteraceae	*Massilia*	2.37±1.40	6.47	100%	13%
Proteobacteria—Uncl. Alteromonadales	*Rheinheimera*	1.89±1.15	4.74	100%	0
Proteobacteria—Rhizobiaceae	*Agrobacterium*	1.85±1.08	4.51	100%	7%
Proteobacteria—Rhodospirillaceae	*Unclassified*	1.61±1.07	5.24	100%	7%
Proteobacteria—Neisseriaceae	*Vogesella*	1.23±0.74	3.04	100%	0
Actinobacteria—Nocardioidaceae	*Nocardioides*	1.09±0.65	2.61	100%	13%
Proteobacteria—Burkholderiales	*Unclassified*	1.07±0.64	2.63	100%	0

定义为在去除潜在的试剂污染物后，至少95%的样品中出现的测序变异情况（ASVs），其平均相对丰度大于1%。Uncl，未分类的。"平均±SD"和"最大"列的单位可能是相对丰度的%。

3.7 宏基因组关联分析的样本量

众所周知，微生物组的数据具有 3 个特点：异质性、肥尾性（极端值的概率很高，不符合正态分布）和零膨胀性（一个分类单元在许多样本中可能是零值）。因此，分析宏基因组数据的统计检验都是非参数的，不依赖于数据的分布。例如，Wilcoxon 秩和检验，也被称为 Mann-Whitney U 检验，通常用于寻找两组之间的差异，即使差异很小，也能产生统计显著的结果。

我们在样本容量估计和统计效力分析方面，对文献 [70] 的方法进行了改进，消除了对初始静态采样空间的限制。理想情况下，为了得到伪相关系数 r^2，一个人需要从大约 40 例样本开始，然后精确地计算出达到 0.8 或更高的统计效力所需的样本大小。

但这已经超过了比较具有强烈和相对同质性差异的样本群体所需的样本量，例如来自类风湿性关节炎患者和健康对照组的牙菌斑，来自 IgA 缺陷者和他们的正常对照的粪便样本。包括更多的样品可能会增加异质性（图 3.8），且两者不一定遵循相同的统计分布。例如，不同的人可能在不同的季节得同一种疾病。因此，另一个在某些方面不同的数据集应该是一个独立的验证，并且也可以发现一些不同的生物标志物。

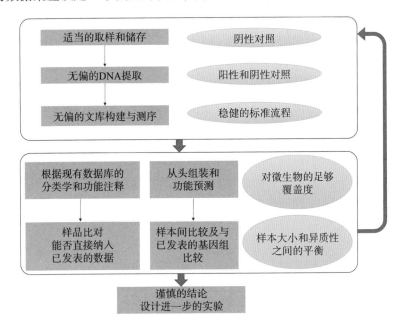

图 3.8　宏基因组研究的流程设计

一个简化的理想的宏基因组研究的流程设计，从样本采集到数据分析和结论。在采集了 20 多个样本后，就可以观察表型的分布，并评估招募的志愿者是否能够回答预期的问题。

我们对某种疾病的认识，可以指导我们选择最关键的身体部位，确定合适的样本量，以及记录或控制相关的表型数据，从而进行宏基因组关联研究（MWAS）。例如，对于类风湿性关节炎患者的牙龈样本、克罗恩病患者的粪便样本或 IgA 缺陷个体的粪便样本，10 例样本即可表现出明显的菌群失调[7,10,48,71-73]。而对于结直肠癌患者，粪便微生物组标志物在几个不同国家的几十个样本（详见第 7 章）就能够很好地收敛。对于代谢性疾病，如肥胖和 2 型糖尿病（更多见于第 6 章），则需要选择极端的表型或更大的样本量。此外，还可以从全民疾病筛查（第 7 章）或没有现代治疗手段的偏远地区获取对照样本。对于大多数医生来说，在复发前检查以前的患者，将是一个更加现实和有临床意义的研究设计。疾病持续时间更长，且曾接受过药物治疗[10,66]，但在停药至少 3 个月后复发的患者，仍然可以展现出相同的疾病标志物，尽管他们的微生物组存在其他差异。因此，药物信息应该被记录下来，并在分析中进行统计学上的调整。

置换多元方差分析（permutational analyses of variances，PERMANOVA）[76,77] 是

一种非参数统计检验，适合于分析表型和问卷对宏基因组数据的潜在影响。虽然这是一个多变量检验，但通常先从单个表型开始分析。针对特定身体部位微生物组设计的调查问卷和其他组学数据，例如怀孕史和激素对子宫颈微生物组、口腔卫生和免疫特征对口腔微生物组，都可以解释微生物组差异的显著部分[14,78]。相反，成人粪便研究队列中，表型数据通常只能解释小于10%的差异[3,79-81]。

对于已知存在显著差异的种群类型的身体部位，如粪便或阴道微生物，我们可以在分析过程中，根据特定的型别对样本进行分类[9,82,83]，虽然这会牺牲一些样本量。据报道，根据肠道型别对人类肠道微生物组项目中粪便数据进行分层，可以提高统计效力[84]。此外，不同性别在微生物组成和免疫反应方面的差异，也意味着在微生物关联分析中分性别分别分析，可能会得出与整合分析不一样的有趣发现[11,83]。

研究人员需要知道影响队列中的重要因素，以便在发现不同群体之间的差异时，能够判断这种差异是由我们正在比较的疾病（或其他条件）造成的，还是由不同群体之间的其他因素（如遗传和生活方式的差异，或种族和地理群体的差异）造成的[50,85,86]。在我们能够合理地建立微生物组模型之前，粗暴的统计控制往往会导致真实信号的丢失。在一项关于大肠癌的研究中，对照组、腺瘤组和癌组的铁结合蛋白和每周红肉摄入量明显不同，这些都是与大肠癌有关的因素，不需要被控制[47]（关于因果关系，详见第6章）。自最后一次月经周期以来的天数（用具体的数值而不是分成几个类别的月经周期），与上部和下部生殖道（从腹腔液、子宫、宫颈到阴道中部）中的许多细菌有关，因此我们不能简单地确定某些细菌与子宫肌瘤和子宫腺肌病的关系[34]。这些特征并不应该仅仅因为统计被忽略，而应该在未来作进一步研究。

除了采用健康人作为疾病对照，采样时还应考虑一些与疾病相关的共同特征。例如，为了更好地理解干燥综合征（Sjögren syndrome，SS）中的口腔微生物组，还需要考虑由其他原因导致的口干人群（口干果然影响了微生物的生存环境）[87]。一项对阻塞性睡眠呼吸暂停患者呼吸道微生物组的研究可能需要考虑志愿者的体脂和心血管健康的差异。同样，如果将已公布的数据作为相关条件的对照，可能会引入很大的异质性[52,88]。在为特定研究领域制定更合适的对照数据之前[6,89]，一种避免表型数据和样本处理差异的方法是始终将疾病样本与同一研究的对照样本进行比较（口干果然影响了微生物的生存环境）[90]。

思考题 3.4

（1）根据双胞胎队列中，同卵和异卵双胞胎粪便微生物的方差分析结果（[50]的表S2，样本量较扩增子测序[90,92]研究更小），我们在下结论时，应该对哪些表型保持谨慎？

（2）我们应该尝试对表型分布有一个直观的认识，例如通过绘图显示在亚洲人的队列中，体质指数（BMI）普遍较低，因此往往不会像欧洲人群的分析那样与肠道菌群呈现出很强的信号。腰臀比、根据电阻率或者扫描得到的肌肉及脂肪量，可能比 BMI 更能反馈腹部肥胖的情况。

（3）基于出生年份和采样年龄之间的差异，在哪些方面会有不同？

（4）你认为根据当前的技术和知识，我们能否更好地研究体育锻炼对肠道菌群的影响？

3.8 总结

生物量较低的样本（即其中的微生物种群很少），可能只是刚刚超过了当前检测的极限，对于样本采集和 DNA 提取过程中的污染更为敏感。除了宏基因组测序，还有多种方法可被用来估计低生物量样本中的微生物数量，并将微生物与对应的功能联系起来。从同一个个体上，采集相关部位的样本，总是一个明智的选择。由于微生物群落是一个复杂的种群，在存储、DNA 提取和测序过程中，要注意避免样本中某些分类单元被低估。在第 5 章中，我们看到，宏基因组组装相比宏基因组检测需要较高的测序覆盖度。针对不同的身体部位，需要调整问卷调查和收集的信息，以免遗漏统计分析过程中的重要信息，并在多个队列中寻找一致性的生物标志物。

Phenotypes	Number of twins	Groups	Sample size	Degree of freedom	Sums of squares	Mean square	F model	Pseudo-R²	P (>F)	P adjusted (BH)
Twin pair number	246	NA	246	122	43.28	0.355	1.196	0.543	0.000	0.002
BMI	249	NA	249	1	0.511	0.511	1.571	0.006	0.002	0.017
Drugs (diabetic tablets)	230	Y	5	1	0.463	0.463	1.428	0.006	0.015	0.083
		N	225							
Has a doctor ever diagnosed or treated you for any of the following conditions?/diabetes	222	Y	10	1	0.429	0.429	1.319	0.006	0.028	0.091
		N	212							
Year of birth	250	NA	250	1	0.424	0.424	1.303	0.005	0.035	0.091
Current location (Geo-clusters)	247	Cluster 1	10	3	1.133	0.378	1.161	0.014	0.037	0.091
		Cluster 2	134							
		Cluster 3	52							
		Cluster 4	51							
Vegetarian or vegan	198	N	179	1	0.420	0.420	1.291	0.007	0.041	0.091
		Y	19							
Age at metagenomic sample	250	NA	250	1	0.417	0.417	1.280	0.005	0.043	0.091
Number of units of alcohol drunk per week	241	1~5 units	88	1	0.389	0.389	1.193	0.005	0.100	0.173
		6~10 units	21							
		11~15 units	56							
		16~20 units	7							
		21~40 units	34							
		40⁺ units	4							
		None	31							
Menopausal status	240	Postmenopausal	174	2	0.734	0.367	1.129	0.009	0.102	0.173
		Premenopausal	38							
		Going through menopause	28							

续表

Phenotypes	Number of twins	Groups	Sample size	Degree of freedom	Sums of squares	Mean square	F model	Pseudo-R²	P (>F)	P adjusted (BH)
Smoking status	249	Smoker	92	1	0.367	0.367	1.126	0.005	0.160	0.247
		Never smoked	157							
Currently, how many minutes per week do you spend walking briskly/gardening vigorously?	196	0	22	1	0.354	0.354	1.096	0.006	0.214	0.303
		≥1	174							
Currently, how many minutes per week do you spend in nonweight bearing activity? e.g., swimming, cycling, yoga, aqua aerobics etc.	188	0	102	1	0.323	0.323	0.998	0.005	0.447	0.559
		≥1	86							
Drugs (Insulin)	230	Y	4	1	0.321	0.321	0.988	0.004	0.460	0.559
		N	226							
Currently, how many minutes per week do you spend on weight-bearing activity? E.g., aerobics, running, dance, football, basketball, racquet sports, etc. (do not include walking or gardening)	191	0	103	1	0.309	0.309	0.953	0.005	0.589	0.667
		≥1	88							
Outdoor sports	108	Y	56	1	0.29	0.29	0.885	0.008	0.798	0.848
		N	52							

对肠道微生物基因图谱（1140万个基因的粪便微生物组[50]）的每个表型的影响进行方差分析。当对每个表型进行一次多变异多变量方法分析（permutational multivariate anaylsis of varianle, PERMANOVA）测试时，使用Benjamini-Hochberg程序进行多重检验控制。表型没有被进行组合分析。

原著参考文献

[1] Vandeputte D, Falony G, Vieira-Silva S, Tito RY, Joossens M, Raes J. Stool consistency is strongly associated with gut microbiota richness and composition, enterotypes and bacterial growth rates. Gut 2016;65:57–62. https://doi.org/10.1136/gutjnl-2015-309618.

[2] Vandeputte D, Kathagen G, D'hoe K, Vieira-Silva S, Valles-Colomer M, Sabino J, et al. Quantitative microbiome profiling links gut community variation to microbial load. Nature 2017;551:507–511. https://doi.org/10.1038/nature24460.

[3] Jie Z, Liang S, Ding Q, Li F, Tang S, Wang D, et al. A transomic cohort as a reference point for promoting a healthy gut microbiome. Med Microecol 2021;8:100039. https://doi.org/10.1016/j.medmic.2021.100039.

[4] Park S, Won DD, Lee BJ, Escobedo D, Esteva A, Aalipour A, et al. A mountable toilet system for personalized health monitoring via the analysis of excreta. Nat Biomed Eng 2020;4:624–635. https://doi.org/10.1038/s41551-020-0534-9.

[5] Methé BA, Nelson KE, Pop M, Creasy HH, Giglio MG, Huttenhower C, et al. A framework for human microbiome research. Nature 2012;486:215–221. https://doi.org/10.1038/nature11209.

[6] Jie Z, Liang S, Ding Q, Li F, Tang S, Sun X, et al. Disease trends in a young Chinese cohort according to fecal metagenome and plasma metabolites. Med Microecol 2021. https://doi.org/10.1016/j.medmic.2021.100037.

[7] He Q, Gao Y, Jie Z, Yu X, Laursen JM, Xiao L, et al. Two distinct metacommunities characterize the gut microbiota in Crohn's disease patients. Gigascience 2017;6:1–11. https://doi.org/10.1093/gigascience/gix050.

[8] Lloyd-Price J, Mahurkar A, Rahnavard G, Crabtree J, Orvis J, Hall AB, et al. Strains, functions and dynamics in the expanded human microbiome project. Nature 2017. https://doi.org/10.1038/nature23889.

[9] Chen C, Hao L, Zhang Z, Tian L, Song L, Zhang X, et al. Dynamics in the vaginocervical microbiome after oral probiotics. J Genet Genomics 2021. https://doi.org/10.1101/2020.06.16.155929.

[10] Zhang X, Zhang D, Jia H, Feng Q, Wang D, Di Liang D, et al. The oral and gut microbiomes are perturbed in rheumatoid arthritis and partly normalized after treatment. Nat Med 2015;21:895–905. https://doi.org/10.1038/nm.3914.

[11] Zhu J, Tian L, Chen P, Han M, Song L, Tong X, et al. Over 50000 metagenomically assembled draft genomes for the human oral microbiome reveal new taxa. Genomics Proteomics Bioinformatics 2021. https://doi.org/10.1016/j.gpb.2021.05.001.

[12] Oh J, Byrd AL, Park M, Kong HH, Segre JA. Temporal stability of the human skin microbiome. Cell 2016;165:854–866. https://doi.org/10.1016/j.cell.2016.04.008.

[13] Fettweis JM, Serrano MG, Brooks JP, Edwards DJ, Girerd PH, Parikh HI, et al. The vaginal microbiome and preterm birth. Nat Med 2019;25:1012–1021. https://doi.org/10.1038/s41591-019-0450-2.

[14] Jie Z, Chen C, Hao L, Li F, Song L, Zhang X, et al. Life history recorded in the vagino-cervical

[15] Li F, Chen C, Wei W, Wang Z, Dai J, Hao L, et al. The metagenome of the female upper reproductive tract. Gigascience 2018;7. https://doi.org/10.1093/gigascience/giy107.

[16] Aagaard K, Ma J, Antony KM, Ganu R, Petrosino J, Versalovic J. The placenta harbors a unique microbiome. Sci Transl Med 2014;6:237ra65. https://doi.org/10.1126/scitranslmed.3008599.

[17] Dickson RP, Erb-Downward JR, Prescott HC, Martinez FJ, Curtis JL, Lama VN, et al. Cell-associated bacteria in the human lung microbiome. Microbiome 2014;2:28. https://doi.org/10.1186/2049-2618-2-28.

[18] Marotz CA, Sanders JG, Zuniga C, Zaramela LS, Knight R, Zengler K. Improving saliva shotgun metagenomics by chemical host DNA depletion. Microbiome 2018;6:42. https://doi.org/10.1186/s40168-018-0426-3.

[19] Castellarin M, Warren RL, Freeman JD, Dreolini L, Krzywinski M, Strauss J, et al. Fusobacterium nucleatum infection is prevalent in human colorectal carcinoma. Genome Res 2012;22:299–306. https://doi.org/10.1101/gr.126516.111.

[20] Kostic AD, Gevers D, Pedamallu CS, Michaud M, Duke F, Earl AM, et al. Genomic analysis identifies association of Fusobacterium with colorectal carcinoma. Genome Res 2012;22:292–298. https://doi.org/10.1101/gr.126573.111.

[21] Riley DR, Sieber KB, Robinson KM, White JR, Ganesan A, Nourbakhsh S, et al. Bacteria-human somatic cell lateral gene transfer is enriched in cancer samples. PLoS Comput Biol 2013;9. https://doi.org/10.1371/journal.pcbi.1003107, e1003107.

[22] Geller LT, Barzily-Rokni M, Danino T, Jonas OH, Shental N, Nejman D, et al. Potential role of intratumor bacteria in mediating tumor resistance to the chemotherapeutic drug gemcitabine. Science 2017;357:1156–1160. https://doi.org/10.1126/science.aah5043.

[23] Poore GD, Kopylova E, Zhu Q, Carpenter C, Fraraccio S, Wandro S, et al. Microbiome analyses of blood and tissues suggest cancer diagnostic approach. Nature 2020;579:567–574. https://doi.org/10.1038/s41586-020-2095-1.

[24] Nejman D, Livyatan I, Fuks G, Gavert N, Zwang Y, Geller LT, et al. The human tumor microbiome is composed of tumor type–specific intracellular bacteria. Science 2020;368:973–980. https://doi.org/10.1126/science.aay9189.

[25] de Goffau MC, Lager S, Sovio U, Gaccioli F, Cook E, Peacock SJ, et al. Human placenta has no microbiome but can contain potential pathogens. Nature 2019;572:329–334. https://doi.org/10.1038/s41586-019-1451-5.

[26] Salter SJ, Cox MJ, Turek EM, Calus ST, Cookson WO, Moffatt MF, et al. Reagent and laboratory contamination can critically impact sequence-based microbiome analyses. BMC Biol 2014;12:87. https://doi.org/10.1186/s12915-014-0087-z.

[27] Sun X, Hu YH, Wang J, Fang C, Li J, Han M, et al. Efficient and stable metabarcoding sequencing data using a DNBSEQ-G400 sequencer validated by comprehensive community analyses. GigaByte 2021. https://doi.org/10.46471/gigabyte.16.

[28] Lax S, Sangwan N, Smith D, Larsen P, Handley KM, Richardson M, et al. Bacterial colonization and succession in a newly opened hospital. Sci Transl Med 2017;9:1–11.

[29] Pidot SJ, Gao W, Buultjens AH, Monk IR, Guerillot R, Carter GP, et al. Increasing tolerance of hospital Enterococcus faecium to handwash alcohols. Sci Transl Med 2018;10:eaar6115. https://doi.org/10.1126/scitranslmed.aar6115.

[30] Mora M, Wink L, Kögler I, Mahnert A, Rettberg P, Schwendner P, et al. Space station conditions are selective but do not alter microbial characteristics relevant to human health. Nat Commun 2019;10:3990. https://doi.org/10.1038/s41467-019-11682-z.

[31] Checinska A, Probst AJ, Vaishampayan P, White JR, Kumar D, Stepanov VG, et al. Microbiomes of the dust particles collected from the international space station and spacecraft assembly facilities. Microbiome 2015;3:50. https://doi.org/10.1186/s40168-015-0116-3.

[32] Lee MD, O'Rourke A, Lorenzi H, Bebout BM, Dupont CL, Everroad RC. Reference guided metagenomics reveals genome-level evidence of potential microbial transmission from the ISS environment to an astronaut's microbiome. IScience 2021;24:102114. https://doi.org/10.1016/j.isci.2021.102114.

[33] Saw JJ, Sivaguru M, Wilson EM, Dong Y, Sanford RA, Fields CJ, et al. In vivo entombment of bacteria and fungi during calcium oxalate, brushite, and struvite urolithiasis. Kidney360 2021;2:298–311. https://doi.org/10.34067/kid.0006942020.

[34] Chen C, Song X, Wei W, Zhong H, Dai J, Lan Z, et al. The microbiota continuum along the female reproductive tract and its relation to uterine-related diseases. Nat Commun 2017;8:875. https://doi.org/10.1038/s41467-017-00901-0.

[35] Chen C, Hao L, Wei W, Li F, Song L, Zhang X, et al. The female urinary microbiota in relation to the reproductive tract microbiota. Gigabyte 2020;2020:1–9. https://doi.org/10.46471/gigabyte.9.

[36] Seferovic MD, Pace RM, Carroll M, Belfort B, Major AM, Chu DM, et al. Visualization of microbes by 16S in situ hybridization in term and preterm placentas without intraamniotic infection. Am J Obstet Gynecol 2019;221:146.e1–146.e23. https://doi.org/10.1016/j.ajog.2019.04.036.

[37] Gosalbes MJ, Llop S, Vallès Y, Moya A, Ballester F, Francino MP. Meconium microbiota types dominated by lactic acid or enteric bacteria are differentially associated with maternal eczema and respiratory problems in infants. Clin Exp Allergy 2013;43:198–211. https://doi.org/10.1111/cea.12063.

[38] Bäckhed F, Roswall J, Peng Y, Feng Q, Jia H, Kovatcheva-Datchary P, et al. Dynamics and stabilization of the human gut microbiome during the first year of life. Cell Host Microbe 2015;17:690–703. https://doi.org/10.1016/j.chom.2015.04.004.

[39] Collado MC, Rautava S, Aakko J, Isolauri E, Salminen S. Human gut colonisation may be initiated in utero by distinct microbial communities in the placenta and amniotic fluid. Sci Rep 2016;6:23129. https://doi.org/10.1038/srep23129.

[40] Wang J, Zheng J, Shi W, Du N, Xu X, Zhang Y, et al. Dysbiosis of maternal and neonatal microbiota associated with gestational diabetes mellitus. Gut 2018. https://doi.org/10.1136/gutjnl-2018-315988. gutjnl-2018-315988.

[41] He Q, Kwok LY, Xi X, Zhong Z, Ma T, Xu H, et al. The meconium microbiota shares more features with the amniotic fluid microbiota than the maternal fecal and vaginal microbiota. Gut Microbes 2020;12:1794266. https://doi.org/10.1080/19490976.2020.1794266.

［42］Ferretti P, Pasolli E, Tett A, Asnicar F, Gorfer V, Fedi S, et al. Mother-to-infant microbial transmission from different body sites shapes the developing infant gut microbiome. Cell Host Microbe 2018;24:133–145.e5. https://doi.org/10.1016/j.chom.2018.06.005.

［43］Massier L, Chakaroun R, Tabei S, Crane A, Didt KD, Fallmann J, et al. Adipose tissue derived bacteria are associated with inflammation in obesity and type 2 diabetes. Gut 2020;69(10):1796–1806. https://doi.org/10.1136/gutjnl-2019-320118.

［44］Anhê FF, Jensen BAH, Varin TV, Servant F, Van Blerk S, Richard D, et al. Type 2 diabetes influences bacterial tissue compartmentalisation in human obesity. Nat Metab 2020;2(3):233–242. https://doi.org/10.1038/s42255-020-0178-9.

［45］Stenkula KG, Erlanson-Albertsson C. Adipose cell size: importance in health and disease. Am J Physiol Integr Comp Physiol 2018;315:R284–295. https://doi.org/10.1152/ajpregu.00257.2017.

［46］Dickson RP, Erb-Downward JR, Martinez FJ, Huffnagle GB. The microbiome and the respiratory tract. Annu Rev Physiol 2016;78:481–504. https://doi.org/10.1146/annurev-physiol-021115-105238.

［47］Feng Q, Liang S, Jia H, Stadlmayr A, Tang L, Lan Z, et al. Gut microbiome development along the colorectal adenoma–carcinoma sequence. Nat Commun 2015;6:6528. https://doi.org/10.1038/ncomms7528.

［48］Moll JM, Myers PN, Zhang C, Eriksen C, Wolf J, Appelberg KS, et al. Gut microbiota perturbation in IgA deficiency is influenced by IgA-autoantibody status. Gastroenterology 2021. https://doi.org/10.1053/j.gastro.2021.02.053.

［49］Zhong H, Ren H, Lu Y, Fang C, Hou G, Yang Z, et al. Distinct gut metagenomics and metaproteomics signatures in prediabetics and treatment-naïve type 2 diabetics. EBioMedicine 2019. https://doi.org/10.1016/j.ebiom.2019.08.048.

［50］Xie H, Guo R, Zhong H, Feng Q, Lan Z, Qin B, et al. Shotgun metagenomics of 250 adult twins reveals genetic and environmental impacts on the gut microbiome. Cell Syst 2016;3:572–584.e3. https://doi.org/10.1016/j.cels.2016.10.004.

［51］David LA, CFC M, Carmody RN, Gootenberg DB, Button JE, Wolfe BE, et al. Diet rapidly and reproducibly alters the human gut microbiome. Nature 2013;505:559–563. https://doi.org/10.1038/nature12820.

［52］Li J, Jia H, Cai X, Zhong H, Feng Q, Sunagawa S, et al. An integrated catalog of reference genes in the human gut microbiome. Nat Biotechnol 2014;32:834–841. https://doi.org/10.1038/nbt.2942.

［53］Bouslimani A, Porto C, Rath CM, Wang M, Guo Y, Gonzalez A, et al. Molecular cartography of the human skin surface in 3D. Proc Natl Acad Sci U S A 2015;112:E2120–2129. https://doi.org/10.1073/pnas.1424409112.

［54］Costea PI, Zeller G, Sunagawa S, Pelletier E, Alberti A, Levenez F, et al. Towards standards for human fecal sample processing in metagenomic studies. Nat Biotechnol 2017. https://doi.org/10.1038/nbt.3960.

［55］Tourlousse DM, Narita K, Miura T, Sakamoto M, Ohashi A, Shiina K, et al. Validation and standardization of DNA extraction and library construction methods for metagenomics-based human fecal microbiome measurements. Microbiome 2021;9:95. https://doi.org/10.1186/s40168-

021-01048-3.

[56] Yang F, Sun J, Luo H, Ren H, Zhou H, Lin Y, et al. Assessment of fecal DNA extraction protocols for metagenomic studies. Gigascience 2020;9(7). https://doi.org/10.1093/gigascience/giaa071.

[57] Fang C, Zhong H, Lin Y, Chen B, Han M, Ren H, et al. Assessment of the cPAS based BGISEQ-500 platform for metagenomic sequencing. Gigascience 2018;7:1–8. https://doi.org/10.1093/gigascience/gix133.

[58] Sender R, Fuchs S, Milo R. Revised estimates for the number of human and bacteria cells in the body. PLoS Biol 2016;14. https://doi.org/10.1371/journal.pbio.1002533, e1002533.

[59] Stephen AM, Cummings JH. The microbial contribution to human faecal mass. J Med Microbiol 1980;13:45–56. https://doi.org/10.1099/00222615-13-1-45.

[60] Lager S, de Goffau MC, Sovio U, Peacock SJ, Parkhill J, Charnock-Jones DS, et al. Detecting eukaryotic microbiota with single-cell sensitivity in human tissue. Microbiome 2018;6:151. https://doi.org/10.1186/s40168-018-0529-x.

[61] Patterson J, Carpenter EJ, Zhu Z, An D, Liang X, Geng C, et al. Impact of sequencing depth and technology on de novo RNA-Seq assembly. BMC Genomics 2019;20:604. https://doi.org/10.1186/s12864-019-5965-x.

[62] Browne PD, Nielsen TK, Kot W, Aggerholm A, Gilbert MTP, Puetz L, et al. GC bias affects genomic and metagenomic reconstructions, underrepresenting GC-poor organisms. Gigascience 2020;9. https://doi.org/10.1093/gigascience/giaa008.

[63] Fang C, Sun X, Fan F, Zhang X, Wang O, Zheng H, et al. High-resolution single-molecule long-fragment rRNA gene amplicon sequencing for uncultured bacterial and fungal communities. bioRxiv 2021. https://doi.org/10.1101/2021.03.29.437457.

[64] Kanehisa M, Goto S, Sato Y, Furumichi M, Tanabe M. KEGG for integration and interpretation of large-scale molecular data sets. Nucleic Acids Res 2012;40:D109–114. https://doi.org/10.1093/nar/gkr988.

[65] Valles-Colomer M, Falony G, Darzi Y, Tigchelaar EF, Wang J, Tito RY, et al. The neuroactive potential of the human gut microbiota in quality of life and depression. Nat Microbiol 2019. https://doi.org/10.1038/s41564-018-0337-x.

[66] Zhu F, Ju Y, Wang W, Wang Q, Guo R, Ma Q, et al. Metagenome-wide association of gut microbiome features for schizophrenia. Nat Commun 2020;11:1612. https://doi.org/10.1038/s41467-020-15457-9.

[67] Koh A, Molinaro A, Ståhlman M, Khan MT, Schmidt C, Manneråas-Holm L, et al. Microbially produced imidazole propionate impairs insulin signaling through mTORC1. Cell 2018;175:947–961.e17. https://doi.org/10.1016/j.cell.2018.09.055.

[68] Rao C, Coyte KZ, Bainter W, Geha RS, Martin CR, Rakoff-Nahoum S. Multi kingdom ecological drivers of microbiota assembly in preterm infants. Nature 2021. https://doi.org/10.1038/s41586-021-03241-8.

[69] Moossavi S, Sepehri S, Robertson B, Bode L, Goruk S, Field CJ, et al. Composition and variation of the human milk microbiota are influenced by maternal and early-life factors. Cell Host Microbe 2019;25:324–335.e4. https://doi.org/10.1016/j.chom.2019.01.011.

[70] Kelly BJ, Gross R, Bittinger K, Sherrill-Mix S, Lewis JD, Collman RG, et al. Power and sample-size estimation for microbiome studies using pairwise distances and PERMANOVA. Bioinformatics 2015;31:2461–2468. https://doi.org/10.1093/bioinformatics/btv183.

[71] Qin J, Li R, Raes J, Arumugam M, Burgdorf KSS, Manichanh C, et al. A human gut microbial gene catalogue established by metagenomic sequencing. Nature 2010;464:59–65. https://doi.org/10.1038/nature08821.

[72] Zou M, Jie Z, Cui B, Wang H, Feng Q, Zou Y, et al. Fecal microbiota transplantation results in bacterial strain displacement in patients with inflammatory bowel diseases. FEBS Open Bio 2019. https://doi.org/10.1002/2211-5463.12744.

[73] Fadlallah J, El Kafsi H, Sterlin D, Juste C, Parizot C, Dorgham K, et al. Microbial ecology perturbation in human IgA deficiency. Sci Transl Med 2018;10. https://doi.org/10.1126/scitranslmed.aan1217, eaan1217.

[74] Wang J, Jia H. Metagenome-wide association studies: fine-mining the microbiome. Nat Rev Microbiol 2016;14:508–522. https://doi.org/10.1038/nrmicro.2016.83.

[75] Liu R, Hong J, Xu X, Feng Q, Zhang D, Gu Y, et al. Gut microbiome and serum metabolome alterations in obesity and after weight-loss intervention. Nat Med 2017;23(7):859–868. https://doi.org/10.1038/nm.4358.

[76] Anderson MJ. A new method for non-parametric multivariate analysis of variance. Austral Ecol 2001. https://doi.org/10.1111/j.1442-9993.2001.01070.pp.x.

[77] Anderson MJ, Walsh Daniel CI. PERMANOVA, ANOSIM, and the Mantel test in the face of heterogeneous dispersions: What null hypothesis are you testing? Ecol Monogr 2013. https://doi.org/10.1890/12-2010.1.

[78] Liu X, Tong X, Zhu J, Liu T, Jie Z, Zou Y, et al. Metagenome-genome-wide association studies reveal human genetic impact on the oral microbiome. Biorxiv 2021. https://doi.org/10.1101/2021.05.06.443017.

[79] Falony G, Joossens M, Vieira-Silva S, Wang J, Darzi Y, Faust K, et al. Population-level analysis of gut microbiome variation. Science 2016;352:560–564. https://doi.org/10.1126/science.aad3503.

[80] Zhernakova A, Kurilshikov A, Bonder MJ, Tigchelaar EF, Schirmer M, Vatanen T, et al. Population-based metagenomics analysis reveals markers for gut microbiome composition and diversity. Science 2016;352:565–569. https://doi.org/10.1126/science.aad3369.

[81] Wang J, Thingholm LB, Skiecevičienė J, Rausch P, Kummen M, Hov JR, et al. Genome-wide association analysis identifies variation in vitamin D receptor and other host factors influencing the gut microbiota. Nat Genet 2016;48:1396–1406. https://doi.org/10.1038/ng.3695.

[82] Gu Y, Wang X, Li J, Zhang Y, Zhong H, Liu R, et al. Analyses of gut microbiota and plasma bile acids enable stratification of patients for antidiabetic treatment. Nat Commun 2017;8:1785. https://doi.org/10.1038/s41467-017-01682-2.

[83] Liu X, Tang S, Zhong H, Tong X, Jie Z, Ding Q, et al. A genome-wide association study for gut metagenome in Chinese adults illuminates complex diseases. Cell Discov 2021;7:9. https://doi.org/10.1038/s41421-020-00239-w.

[84] Mattiello F, Verbist B, Faust K, Raes J, Shannon WD, Bijnens L, et al. A web application

for sample size and power calculation in case-control microbiome studies. Bioinformatics 2016;32:2038–2040. https://doi.org/10.1093/bioinformatics/btw099.

[85] He Y, Wu W, Zheng HM, Li P, McDonald D, Sheng H-F, et al. Regional variation limits applications of healthy gut microbiome reference ranges and disease models. Nat Med 2018;24:1532–1535. https://doi.org/10.1038/s41591-018-0164-x.

[86] Deschasaux M, Bouter KE, Prodan A, Levin E, Groen AK, Herrema H, et al. Depicting the composition of gut microbiota in a population with varied ethnic origins but shared geography. Nat Med 2018;24:1526–1531. https://doi.org/10.1038/s41591-018-0160-1.

[87] AlmståhI A, Wikström M, Stenberg I, Jakobsson A, Fagerberg-Mohlin B. Oral microbiota associated with hyposalivation of different origins. Oral Microbiol Immunol 2003;18:1–8.

[88] Karlsson FH, Tremaroli V, Nookaew I, Bergström G, Behre CJ, Fagerberg B, et al. Gut metagenome in European women with normal, impaired and diabetic glucose control. Nature 2013;498:99–103. https://doi.org/10.1038/nature12198.

[89] Jie Z, Xia H, Zhong SL, Feng Q, Li S, Liang S, et al. The gut microbiome in atherosclerotic cardiovascular disease. Nat Commun 2017;8:845. https://doi.org/10.1038/s41467-017-00900-1.

[90] Kachroo N, Lange D, Penniston KL, Stern J, Tasian G, Bajic P, et al. Standardization of microbiome studies for urolithiasis: an international consensus agreement. Nat Rev Urol 2021;18:303–311. https://doi.org/10.1038/s41585-021-00450-8.

[91] Goodrich JK, Waters JL, Poole AC, Sutter JL, Koren O, Blekhman R, et al. Human genetics shape the gut microbiome. Cell 2014;159(4):789–799. https://doi.org/10.1016/j.cell.2014.09.053.

[92] Goodrich JK, Davenport ER, Beaumont M, Jackson MA, Knight R, Ober C, et al. Genetic determinants of the gut microbiome in UK twins. Cell Host Microbe 2016;19(5):731–743. https://doi.org/10.1016/j.chom.2016.04.017.

第 4 章

人体中的流行病学

摘　要：本章着重于确定特定微生物组的来源，无论是来自环境、家庭成员抑或是同一个人的其他部位。第 3 章的知识将用于各种样本的研究。许多细菌可以进入淋巴结或进入体液循环。真菌在出现其他严重症状之前可以潜伏在肠道内。对于类风湿性关节炎患者，滑液中含有在口腔、肺部或粪便微生物组中鉴定出的细菌 DNA。动脉粥样硬化斑块中含有多种口腔微生物。代谢物的流动以及组织内的免疫细胞，也可能为何处寻找微生物罪魁祸首提供线索。了解病源是长期有效管理微生物组相关疾病的先决条件。

关键词：流行病学，类风湿性关节炎，心血管疾病，血源性传播，菌血症，隐匿性败血症，淋巴引流，循环，真菌群，代谢组学

4.1　和新型冠状病毒疫情类比

新冠大流行，提高了对流行病学和传播跟踪的认识。人类微生物群落不仅与日常和环境交换微生物，还包含身体内部的微生物运动，以及有时和人体细胞的互作。从生态学的角度来看（第 2 章），特定身体部位的微生物群落，是由多个源种群和个体相关的因素共同作用，形成具有稳定丰度的微生物种群。控制菌群来源和改善局部条件，将有助于实现更有效的治疗方式。

我们在第 3 章讨论了样本采集，以及在采集和追踪与疾病相关的微生物时，适用于各类样本的一些原则。目前，成功且全面研究的案例还很少，但我们期待在未来会有更多可以讨论的案例。

思考题 4.1

假设你去一个偏远的村庄旅游，在你自己的微生物群落中能包含这些新的微生物之前，那些当地人及环境的样本你计划如何研究？你预期在多少天之后，会看到你的皮肤、口腔和粪便样本出现差异？

思考题 4.2

结合第 3 章内容以及新型冠状病毒感染疫情的经验，认识到人类会不可避免地对周围环境带来微生物污染。本节思考题是为第 5 章的分类主题做一个预热。

基于 HMP 的数据[1]，图 4.1 展示了关于纽约地铁微生物的一项研究。我们现在还可以利用哪些数据，更好地将微生物数据归类到不同的身体部位上？在分析某个特定地铁站的数据时，我们是否要考虑不同人种的数据？在刚被洪水侵袭的地铁站中，作者检测到了传统上认为是海洋来源的假交替单胞菌属（*Pseudoalteromonas*），根据目前数据，这一结论是否有更新？

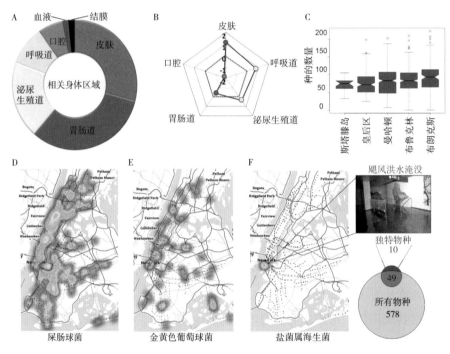

图 4.1 在纽约地铁采样的样本的分类群多样性和可能的人体部位

根据美国人类肠道微生物组项目协会数据集，我们对检测到的细菌进行了注释，并得到了它们的相对丰度。A. 在与 HMP 数据集匹配的 67 个 PathoMap 物种中，胃肠道（蓝色）、皮肤（绿色）和泌尿生殖道（白色）的比例最大。整个圆圈代表了 67 种细菌中的 100%，每种颜色的大小代表了每种细菌的比例。B. 为了解释 HMP 数据库中的比例，我们计算了每个类别中实际观察到的物种数与理论预期发现的物种数的 log2 比值，发现皮肤是地铁系统中最常见的细菌来源。C. 每个行政区发现的物种数量的箱线图。每个方框的中间线代表中位数，每个方框的顶部和底部分别显示第 75 和 25 百分位数。两个方框之间的缺口显示两组之间有显著差异（95%的置信区间）。D. 和 E. 纽约热图显示屎肠球菌（D）和金黄色葡萄球菌（E）的分布密度。小红点表示在该位置发现了一个完全相同的 *mecA* 基因。F. 对飓风桑迪期间被洪水淹没的地铁站的分析（上图为车站）。维恩图展示了来自该站的 10 个物种的独特组合，这些物种在纽约市的任何其他车站或地区都没有出现，而另外 52 个物种与地铁系统中的 627 个物种组合重叠。（A~F）整个纽约城市运输局地铁系统共有 468 个车站，在 2013 年夏季期间进行了三次擦拭采样，并在 2014 年为培养和测试以及回应审查人员的要求，又采集了一些额外的样本。每个车站擦拭两个不同的表面，列车内擦拭一个表面。从十字转门和紧急出口、地铁卡亭、木制和金属长椅、楼梯扶手和垃垃圾桶中采集样本。由于在这些特定地点人与人之间的频繁接触，每个车站的十字转门和报刊亭都得到了优先安排。在火车上，车门、杆子、扶手和座椅都擦拭过。来源：Cell Syst, 2015, 1:1-15. doi:10.1016/j.cels.2015.01.001.

4.2 婴儿肠道共生菌的来源

当婴儿出现腹泻症状，如果能够恰巧采集到适合进行高通量宏基因组测序样本，我们能否判断腹泻的原因是婴儿摄入的食物或水中的某种物质[2]？是否更换了奶粉（除了营养物质，还包含牛奶中的细菌孢子）？是不是直接的母乳喂养比吸出来再喂养更有利[3]？我们是不是应该更多地关注母亲的肠道微生物？除了通过环境传播，一些母亲的肠道微生物可能会通过乳腺进入母乳（根据 PCR，每毫升包含 10^6 个细菌，10^5 个真菌[4,5]）。婴儿经常会把带有微生物的乳汁吐回给母亲，因此婴儿的微生物也应纳入循环（图 4.2）。口腔–肠道的传播在婴儿中比成人中更普遍[6]。用扩增子测序法分析母乳和乳晕皮肤（乳环）微生物群时，发现对于母乳喂养比例超过 75% 的婴儿，母乳中的细菌占婴儿肠道微生物群的 30%，但在第 1 个月后，母乳对肠道微生物群的贡献率降低到低于 10%[7]。

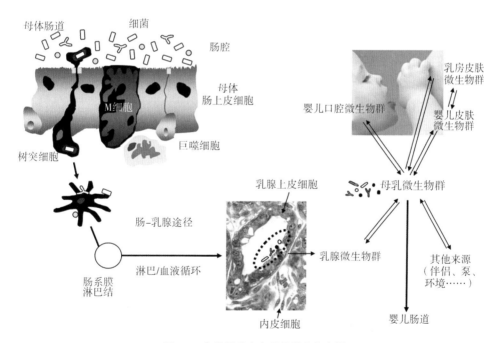

图 4.2 人类母乳中出现的微生物来源

用模型来解释一些母体细菌菌株是如何通过母乳喂养传播到婴儿的肠道的。来源：Adv Nutr, 2014, 5:779-784. https://doi.org/10.3945/an.114.007229.

4.3 共生微生物的异位存续

许多微生物只要待在其应待的地方，就是无害的。然而，在某些罕见情况下（将在第 8 章深入研究），或由于在正常环境下的某个过程被逆转，这些微生物可能会出现并在错误的地方生长。

克罗恩病和溃疡性结肠炎患者的唾液细菌在无菌小鼠肠道内的定植已被证明能够引发炎症[8]。宏基因关联分析（MWAS）鉴定的一些与人类疾病相关的生物标志物可能来自口腔，除了年龄和其他生理条件之外，它们在肠道的存续可能受到遗传因素的影响[9,10]。

在癌症小鼠模型中，研究表明肠道微生物会迁移到次级淋巴结中，从而激活免疫反应和促进治疗效果（第 7 章，表 7.2）。

缺乏丁酸导致的肠道屏障受损与很多疾病有关，而炎症通常归因到了脂多糖。血液中可以检测到细菌 DNA[11-13]，但是我们不清楚这些细菌是否在健康的成年人体内存活（亚临床的菌血症）。有些已知的病原体可以隐藏在免疫细胞内，例如肺炎链球菌能够在脾巨噬细胞内繁殖[14]。产脓链球菌在淋巴管的多个淋巴结之间转移进入血液时，已经被证明位于细胞外[15]。在接受骨髓移植（allo-HCT）的患者中，在发生血液感染（败血症）之前，万古霉素耐药肠球菌（*vancomycin-resistant Enterococcus*）已经在肠道大量存在[16]。在引起血液感染之前，真菌念珠菌（*Candida* spp.）在骨髓移植患者的肠道中增殖（图 4.3，回顾第 1 章微生物细胞数量的问题）。

另外，研究发现无菌鼠的血 - 脑脊液屏障受损，可通过补充丁酸、酪丁酸梭菌 *Clostridium tyrobutyricum* 或多形拟杆菌 *Bacteroides thetaiotaomicron*（产生乙酸和丁酸两种其他的主要短链脂肪酸）来改善[17]。激素周期也会影响血 - 脑脊液屏障的稳定性，其中雌激素有保护作用[18]（关于月经周期，详见第 8 章）。关于脑肠轴，还有许多问题值得探讨。

4.4 病灶部位菌群

4.4.1 类风湿关节炎

粪便、唾液和牙菌斑的微生物群落和类风湿关节炎的关系已被研究过，微生物很可能促成了免疫失调[19,20]。口腔和粪便样本中唾液乳杆菌的相对丰度呈相关性，类风湿关节炎患者中富集的菌株可能与益生的唾液乳杆菌有所不同。据报道，口干症患者富含乳酸菌[19]。吸烟者口腔中的普雷沃菌属和韦荣氏球菌（*Veillonella*）的丰度更高[21]。

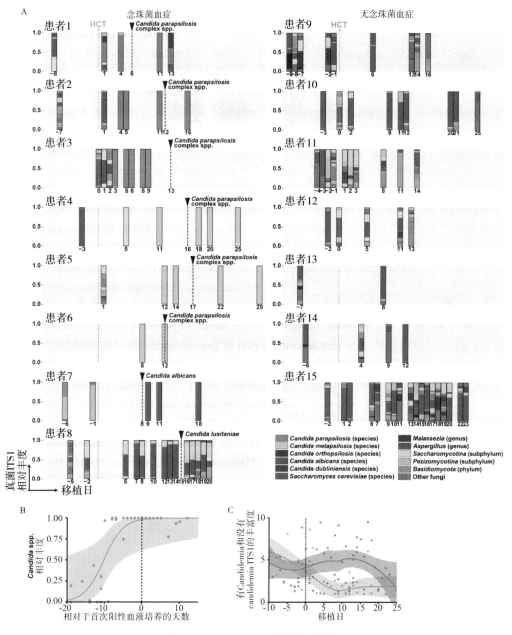

图 4.3　allo-HCT 患者菌群动态研究

A. 所有患者粪便菌群物种水平分类（患者共采集 2~18 个样本，平均 7 个）左边为有念珠菌的，右边为无念珠菌的。其他常见的物种，例如念珠菌属和酿酒酵母属，都有单独的颜色编码。灰色方框表示移植第 10 天至第 30 天，灰色虚线表示移植日。每个条形图下面的数字表示抽样的日期。黑色的虚线和箭头表示念珠菌血症组中第一次真菌血液感染的日期。B. 从患者 1 到患者 7，每份粪便样本（$n=37$）中致病性念珠菌的总相对丰度定量；实线表示动态趋势，阴影部分表示 95% 的置信区间。C. 用逆辛普森指数测定每个样本真菌群的 α-多样性。红点线：白血病组（$n=51$）；绿松石点线：非白血病组（$n=57$）。来源：Nat Med, 2020, 26:59-64. https://doi.org/10.1038/s41591-019-0709-7.

肺部微生物虽然与吸烟对类风湿关节炎的影响和类风湿关节炎死亡率有关，但由于需要侵入性的肺泡灌洗液才能采样，因此很少被研究。一项对20名新发类风湿关节炎患者、10名结节病患者和28名健康对照组的研究发现，韦荣氏球菌与抗环瓜氨酸肽（CCP）的IgA和抗类风湿因子（RF）的IgA相关，这与另一队列的口腔微生物组结果一致[19]。一个未分类的草酸杆菌科细菌与DAS28（疾病活动评分）呈负相关，使人想起一个接近于草酸杆菌的细菌，出现在另一个比较类风湿关节炎患者和对照组的粪便微生物组的独立研究中[22]。文献中也有滑膜液中存在细菌的证据。对培养阴性的滑膜液样本中特定的牙科细菌进行PCR检测，经常能检测到细菌，但在白细胞则为PCR阴性[23]。例如，难治性类风湿关节炎患者中，19/19的牙菌斑、14/19的血清和17/19的关节液中，都检测到中间型普雷沃菌；15/19的牙菌斑、8/19的血清和11/19的关节液样本中检测到牙龈卟啉单胞菌；4/19的牙菌斑、0/19的血清和3/19的关节液样本中检测到伴放线凝聚杆菌[23]。对110例类风湿关节炎患者和42例骨性关节炎（OA）患者的关节液标本进行16S rRNA基因扩增测序，发现殊异韦荣球菌（*Veillonella dispar*）、副流感嗜血杆菌（*Haemophilus parainfluenzae*）、普雷活特氏菌、奇异菌和密螺施体（*Trepanema amylovorum*）在类风湿关节炎患者中相对丰度更高，而粪便拟杆菌属在OA患者中相对丰度更高[24]。这些类风湿关节炎患者或对照组中的滑膜液中的富集细菌（或同属的另一种细菌）都曾被报道为类风湿关节炎患者肠道或口腔微生物组中的重要生物标志物[19,20,25,26]。根据标准的微生物学方法，所有这些滑液样本均为传统培养阴性。这些研究还收集了滑膜组织样本，结果与滑膜液中细菌生物标志物不一致[24]。

伴放线凝聚杆菌（*Aggregatibacter actinomycetemcomitans*），而不是其他更常见的牙周炎病原体（普雷沃菌，福赛斯坦纳菌，齿状密螺旋体，具核梭菌，微小帕维单胞菌，中间普雷沃菌），通过其表达的成孔毒素白细胞毒素A（LtxA）在中性粒细胞中引起超瓜氨酸化（到蛋白质）[26]。抗LtxA抗体在类风湿关节炎患者和牙周炎患者中富集，并与抗环瓜氨酸蛋白抗体（ACPA）阳性和类风湿因子阳性类风湿关节炎[26]重叠。伴放线凝聚杆菌（更名自放线杆菌）、卟啉假单胞菌和中间普雷沃菌的内毒素在体外诱导破骨细胞分化的实验中也呈阳性[27]，可能导致骨质吸收，这是类风湿关节炎病情恶化的因素之一。

除了Th17细胞活化[28]，最近的单细胞研究表明，在牙周炎患者的口腔黏膜中，成纤维细胞招募了中性粒细胞和其他免疫细胞[29]。白细胞大多数是中性粒细胞，常规血液检测结果显示，中性粒细胞与唾液乳杆菌、韦荣氏球菌等类风湿关节炎富集微生物呈正相关，而与富含对照菌如乳酸乳球菌呈负相关[19]（图4.4）。

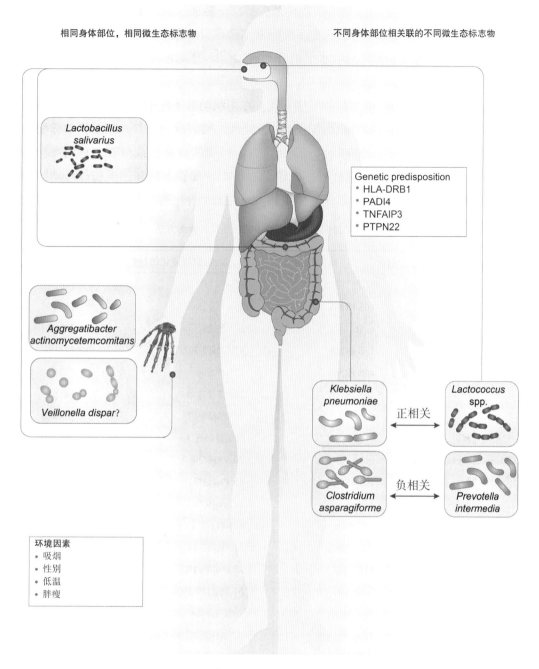

图 4.4　类风湿关节炎中的微生物组紊乱

口腔和粪便样本为非侵入性采集,而来自关节或肺部的样本则更难获取。这也适用于免疫细胞群体的问题。滑液结果尚未进行鸟枪法宏基因组学分析,并且尚未与同一患者口腔宏基因组数据相匹配。

4.4.2　心血管疾病

肝硬化患者的粪便中含有过多的可能来自口腔的细菌,在停止使用质子泵抑制剂

（PPI，如奥美拉唑）后，这些细菌数会减少[30-33]。然而，我们不清楚这些细菌是否也存在于肝脏中。使用 PPI 可以让更多的唾液细菌在胃中存活，并且增加粪便中链球菌的相对丰度，这与血清胃泌素水平的升高有关[33]。牙周治疗在一组肝硬化患者中取得了良好的效果[34]。

很多心血管事件，可能和口腔卫生有关。在粥状动脉硬化中，已经报道了很多相关的细菌（表 4.1）[35]，这些细菌都可能是口腔微生物的成员[77,78]。与粥状动脉硬化相关的肺炎克雷伯菌（*Klebsiella pneumoniae*），可能来自肠道或口腔，它们参与了脂多糖和三甲胺的产生，其在肝脏中代谢成 TMAO，而脂多糖和三甲胺都与动脉粥样硬化心血管疾病有关[79,80]。建议对微生物组进行更个性化的监测（第 8 章）。

表 4.1 动脉粥样硬化斑块中的细菌及其检测方法

斑块样本检测到该菌的比例	检测平台	动脉粥样硬化斑块的细菌	参考文献
Aggregatibacter actinomycetemcomitans [门：Proteobacteria]	PCR	71.4%[5/7]	[36]
	16S rRNA	66.67%[28/42]	[37,39]
	mAb	17%[5/29]	[39,40]
	16S rRNA	21.87%[7/32]	[41]
	16S rRNA	18%[9/50]	[42]
	16S rRNA	25.9[7/27]	[43]
	RT-PCR	46.2%[18/39]	[44,45]
	16S rRNA	294%[15/51]	[46,47]
Chiamydiae pneumoniae [门：Chlamydiae]	mAb	20.6%[6/29]	[39,40]
	16S rRNA	35.4%[11/31]	[39,41]
	16S rRNA	18%%[9/50]	[39,42]
	ICC/PCR	48%[11/23]	[48]
	16S rRNA	51.5%[17/33]	[49]
	MIF lgA	32.6%[63/193]	[50]
	MIF lgA	61.7%[119/193]	[50]
	16S rRNA	26%[12/46]	[51]
	PCR	42%[102/241 sections(10 samples)]	[52]
	PCR	69%[11/16]	[53]
	lmmunofluorescence	79%[71/90]	[54]
	PCR	70%[42/60]	[55]
	lgG antibody	61.7%[50/81]	[56]

续表

斑块样本检测到该菌的比例	检测平台	动脉粥样硬化斑块的细菌	参考文献
Campylobacter rectus [门: Proteobacteria]	16S rRNA	9.52%[4/42]	[37,38,57]
	PCR	11.7%[6/51]	[44,46]
	16S rRNA	21.51%[11/51]	[44,46,58]
	16S rRNA	15.7%[8/51]	[59]
	16S rRNA	21.51%[11/51]	[43]
Enterobacter hormaechei [门: Proteobacteria]	16S rRNA	50%[134/268]	[60]
	16S rRNA	40%[2/5]	[61]
	16S rRNA	54.76%[23/42]	[37,38]
	PCR	15.6%[8/51]	[57]
	16S rRNA	27.45%[14/51]	[59]
Fusobacterium nucleatum [门: Fusobacteria]	16S rRNA	50%[21/42]	[37,38]
	Monoclonal antibody	34%[10/29]	[57]
	PCR	21%[4/19]	[59]
Fusobacterium nucleatum [门: Fusobacteria]	—	—	[63-65]
Helicobacter phlori [门: Fusobacteria]	lgA	55.4%[107/193]	[50]
	lgM	44.6%[86/193]	[50]
	16S rRNA	37%[17/46]	[51]
	lHC	57.8%[22/38]	[66]
	PCR	92.16%[47/51]	[67]
	lgG	67.9%[55/81]	[56]
Mycoplasma pneumoniae [门: Tenericutes]	Seropositivity	14%[396]	[68]
	—	—	[69]
Porphyromonas endodontalis [门: Bacteriodetes]	—	—	[70]
Porphyromonas gingicalis [门: Bacteriodetes]	16S rRNA	78.57%[33/42]	[37,38]
	PCR	71.43%[5/7]	[36]
	16S rRNA	67%[134]	[60]
	mAb	52%[15/29]	[39,40]
	16S rRNA	22.27%[6/22]	[39,41]
	16S rRNA	26%[13/50]	[39,42]
	PCR	47.4%[9/19]	[62]
	PCR	51%[27/53]	[71,72]
	PCR	43.1%[22/51]	[57]

续表

斑块样本检测到该菌的比例	检测平台	动脉粥样硬化斑块的细菌	参考文献
Porphyromonas gingicalis [门：Bacteriodetes]	16S rRNA	45.1%[23/51]	[44,45]
	16S rRNA	21.6%[11/51]	[44,46,58]
	RT-PCR	53.8%[21/39]	[44,45]
	16S rRNA	45.1%[23/51]	[59]
	16S rRNA	7.4%[2/27]	[43]
Prevotella intermeda [门：Bacteroidetes]	mAb	41%[12/29]	[39,40]
	16S rRNA	9.37%[3/32]	[39,41]
	16S rRNA	14%[7/50]	[39,42]
	PCR	21%[4/19]	[61]
	PCR	15%[8/53]	[72,73]
	PCR	19.6%[10/51]	[57]
	RT-PCR	79.3%[23/29]	[44,45]
	PCR	71.43%[5/7]	[36]
	16S rRNA	3.7%[1/27]	[43]
	PCR	15.6%[5/51]	[57]
	RT-PCR	17.9[7/39]	[44,45]
Pseudomonas aeruginosa [门：Proteobacteria]	16S rRNA	40%[6/15]	[74][a]
Pseudomonas luteola [门：Proteobacteria]	16S rRNA	100%[15/15]	[75]
Streptococcus gordonii	PCR	19.4%[-]	[43][b]
Streptococcus mitis	PCR	19.4%[-]	
Streptococcus mutans	PCR	74.1%[20/27]	
Streptococcus oralis	PCR	3.7%[1/27]	
Streptococcus sanguinis [门：Spirochaet]	PCR	25.9%[7/27]	
Treponema denticola [门：Spirochaetes]	PCR	43%[23/53]	[44,71]
	16S rRNA	44.4%[12/27]	[43]
	PCR	35.2%[18/51]	[57]
	16S rRNA	49.01%[25/51]	[44,46]
	16S rRNA	27.4%[14/51]	[44,46,47]
	16S rRNA	23.1%[6/26]	[43,47]
	16S rRNA	49.01%[25/51]	[59]

续表

斑块样本检测到该菌的比例	检测平台	动脉粥样硬化斑块的细菌	参考文献
Tannerella forsythia [门：Bacteriodetes]	16S rRNA	61.9%[26/42]	[37]
	PCR	100%[7/7]	[36]
	mAb	34%[10/29]	[39,40]
	16S rRNA	30%[15/50]	[39,42]
	PCR	10.5%[2/19]	[62]
	PCR	19.6%[10/51]	[57]
	16S rRNA	5.9%[3/51]	[38,44,46]
	RT-PCR	25.6%[10/39]	[44,45]
Veillonella [门：Firmicutes]	16S rRNA	10%[2/20]	[76]
	16S rRNA	100%[13/13]	[75]

经尾静脉注射后，在小鼠胎盘中发现了人唾液微生物组和龈下微生物组成员[81]。除了早产和足月死胎的人类病例外，梭杆菌属也与先兆子痫有关。接受先兆子痫患者粪便移植的小鼠表现为胎盘内含有梭杆菌，促炎细胞因子和趋化因子如 IL-6、IL-1β、CCL3、CCL4 的表达升高[82]。

在健康阴道中，卷曲乳杆菌是优势菌，与睾酮等激素有关[83,84]，粪便微生物组中卷曲乳杆菌的丰度与阴道内卷曲乳杆菌的丰度相关[85]。然而，在动脉粥样硬化患者的粪便微生物组中，卷曲乳杆菌和其他乳酸杆菌的丰度更高（第 2 章，图 2.14）；使用他汀类药物的女性，与那些没有使用他汀类药物的女性相比，无论是胆固醇正常的女性还是胆固醇偏高但没有使用他汀类药物的女性[86]，阴道菌群以卷曲乳杆菌为主，并且占比更高。他汀类药物对卷曲乳杆菌的调节作用及其在人体内分布的性别差异有待进一步研究。

思考题 4.3

根据前一章提到的微生物的理论和实践知识，你想研究肺部的什么问题（图 4.5）？你会如何设计采样和其他信息的收集？

4.5 微生物组中的跨界相互作用

共生菌通常偏好中性 pH 和体温，而真菌能够在更低的 pH、干燥和不太合适的环境中生存。遗憾的是，针对细菌的扩增子测序无法检测到真菌，针对真菌的扩增子测序无法检测到细菌（第 3 章，图 3.4）。根据常规的宏基因组测序的测序深度（如果样本有被适当地提取），健康人的肠道、口腔和阴道中，真菌的丰度很低[85,87,88]，而真菌在皮肤中的分布更为广泛[89]。身体各部位都可能因为感染或者其他原因出现真菌失衡（图4.6）。如果有更多数据，我们可能可以在真菌丰度还不高时就预测它即将过度增殖，以及与之共存的细菌是否还能守住自己的阵地[90]。

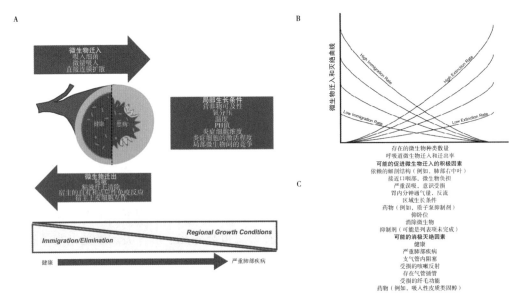

图 4.5 呼吸道微生物的生态学建模

A. 肺部微生物的组成由3个因素决定：来自口腔和上呼吸道的微生物侵入，局部的微生物清除以及组成微生物的相对繁殖速度。在健康的肺部，微生物组成主要受侵入和清除的影响，在患有严重肺部疾病的肺部环境下，侵入和清除都受到损伤，微生物主要受局部生长条件的制约，而同一物种可能会出现不同的菌株。B. 肺部生物地理学的岛屿模型。对于呼吸道中的一个特定部位来说，健康者的菌群丰度由侵入和清除因素共同决定。C. 推测的肺微生物群的促进侵入因素和负向清除因素。来源：A. Annu Rev Physiol, 2016, 78:481-504. B. Proc Natl Acad Sci USA, 2014, 111:13145-13150. C. Lancet Respir Med, 2014, 2:238-246. https://doi.org/10.1016/S2213-2600(14)70028-1.

4.6 其他显示微生物存在差异的组学数据

除了对微生物组进行测序，其他的组学数据也能提供关于哪些情况可能异常的有用信息。在猪等动物模型中，已经系统地研究了每个器官代谢物的进出（表4.2和表4.3）[94]。代谢组学技术也正在发展为单细胞水平的测量[95]。每个器官内的微生物，连同宿主酶，可能影响了特定代谢物的水平，如氨基酸、短链脂肪酸。这些代谢物能进一步作用于微生物，产生对微生物的差异性生长或抑制[96]。一个有些牵强的类比是，动脉、静脉和淋巴循环就像1854年伦敦发生的霍乱疫情中的污水系统，找到微生物组的罪魁祸首是很重要的。

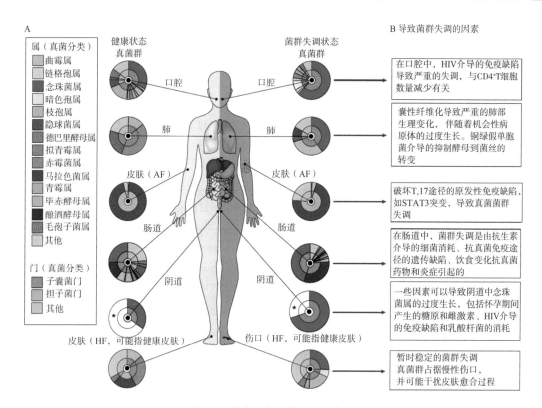

图 4.6 健康和失调状态下的菌群

A. 在内稳态,多样的真菌种群生活在人体屏障表面,例如口腔、肺部、皮肤及阴道。饼图代表门水平和基因水平观察到的相对丰度。值得注意的是,由于缺少基于测序的关于疾病条件的研究,关于阴道的数据是基于细菌培养研究得到的估计值(用*标记),其他指测序得出的相对丰度小于5%。在疾病状态,这些真菌种群的平衡会被打破。在艾滋病毒感染者的口腔和阴道、囊性纤维化患者的肺、原发性免疫缺陷和慢性创伤患者的皮肤以及克罗恩病患者的肠道中都可以观察到菌群失调状态下的真菌菌群。B. 导致不同屏障表面真菌菌群失调的因素。来源:Nat Rev Immunol, 2017. https://doi.org/10.1038/nri.2017.55.

表 4.2 猪各器官的代谢物产生与消耗

器官	代表性发现	关键证据
肝脏	清除不饱和脂肪酸 生产氨基酸	与最丰富的饱和脂肪酸相比,微生物代谢与油酸酯(C18:1)和亚油酸酯(C18:2)的吸收更相关,并对三羧酸循环(TCA)的贡献更大 释放显著数量的氨基酸
小肠	消耗果糖和氨基酸	在任何器官,对于果糖和氨基酸的最大绝对摄入量
胰腺	生产三羧酸的中间产物	枸橼酸盐、酮戊二酸盐、琥珀酸盐、富马酸盐和苹果酸盐的显著释放
脾	生产核苷	胞苷、脱氧胞苷、脱氧尿苷、鸟苷、肌苷、胸苷、尿苷和黄嘌呤的显著释放
大脑	生产不饱和甚长链脂肪酸 生产乙酸	C22:1, C22:2, C22:3, C22:4, C22:5, C22:6, C24:1, C24:2, C24:3, C24:4, C24:5 和 C24:5 的显著释放,在静脉血中的乙酸盐含量增加 2 倍以上

续表

器官	代表性发现	关键证据
腿部肌肉	消耗短链酰基肉碱 生产长链酰基肉碱	摄取 C2：0，C3：0，C4：0，C5：0，和 C5：1 肉碱 C8：0，C10：0，C12：0，C12：1，C14：1，C14：2，C16：0、C16：1，C18：1，C18：2 和 C20：4 肉碱的显著释放
心脏	消耗长链脂肪酸	C16：0，C16：1，C18：0，C18：1，C18：2，C20：1，C20：2，C22：4，C24：0，和 C24：1 的显著消耗
肺	生产不饱和长链脂肪酸	C20：2，C22：4 的显著释放
肾		

表 4.3 猪的每个器官产生和消耗最多的三种代谢产物

器官	产生		消耗	
	代谢物	静脉/动脉比值的对数（以2为底）	代谢物	静脉/动脉比值的对数（以2为底）
肝	谷氨酸	0.64±0.11	胆汁酸 (5)	−2.89±0.19
	三乙醇胺	0.49±0.17	苯丙酸 (2)	−2.29±0.12
	乙酰乙酸	0.38±0.09	短链脂肪酸 (3)	−2.02±0.83
门静脉（肠道）	胆汁酸 (6)	3.28±0.21	2-甲基马尿酸	−0.69±0.15
	苯丙酸 (2)	2.84±0.32	葡萄糖	−0.31±0.05
	短链脂肪酸 (3)	2.82±1.15	谷氨酰胺	−0.28±0.02
结肠	短链脂肪酸 (3)	4.65±1.21	2-甲基马尿酸	−0.60±0.21
	石胆酸	4.04±1.10	5-羟赖氨酸	−0.41±0.05
	苯丙酸 (2)	3.42±0.48	葡萄糖	−0.39±0.05
胰腺	黄嘌呤	1.05±0.26	5-羟赖氨酸	−0.79±0.19
	辛酸甘氨酸	0.51±0.09	N-carbamoylsarcosine	−0.39±0.09
	三羧酸循环中间体 (5)	0.36±0.17	Amino acids (8)	−0.36±0.01
脾脏	O-磷酸乙醇胺	1.11±0.22	Adenosine	−0.61±0.14
	核苷 (9)	0.52±0.03	Dihydroxymandelic acid	−0.33±0.07
	C22 和 C24 非常长链脂肪酸 (11)	0.35±0.008	C5 酰基肉碱	−0.26±0.02
头部（脑）	辛弗林	1.89±0.68	Dihydroxymandelic acid	−0.38±0.12
	葡萄糖酸内酯和葡萄糖酸	1.66±0.03	2-甲基马尿酸	−0.34±0.11
	乙酸	1.46±0.39	谷氨酰胺	−0.30±0.09
腿（肌肉）	次黄嘌呤	0.69±0.12	谷氨酰胺	−1.41±0.33
	支链羟基酸 (2)	0.65±0.12	Ketone bodies (2)	−0.58±0.14
	中链和长链酰基肉碱 (11)	0.57±0.02	短链酰基肉碱 (5)	−0.36±0.05
肺部	2-苯丙酸	0.48±0.14	5-Keto-d-gluconic acid	−0.31±0.07
	顺乌头酸	0.26±0.03	犬尿氨酸	−0.22±0.03
	C22:0 和 C24:0 脂肪酸	0.24±0.02	3-羟基蒽醌酸	−0.17±0.02
肾	甘氨酰胺	1.87±0.12	N-甲酰-L-蛋氨酸	−2.66±0.32
	丝氨酸	0.73±0.12	中链酰基肉碱 (4)	−2.61±0.19
	尿囊素	0.53±0.12	N-乙酰氨基酸 (9)	−1.27±0.87

续表

器官	产生		消耗	
	代谢物	静脉/动脉比值的对数（以2为底）	代谢物	静脉/动脉比值的对数（以2为底）
心脏	次黄嘌呤	0.34±0.11	3-苯丙酸	−0.71±0.24
	谷氨酸	0.26±0.04	不饱和脂肪酸(11)	−0.53±0.06
	生物素	0.25±0.06	羟吲哚乙酸	−0.47±0.12
耳朵(皮肤)	鸟嘌呤	0.82±0.19	羟基马尿酸	−0.35±0.08
	牛磺酸	0.53±0.08	吲哚代谢物（2）	−0.23±0.02
	长链酰基肉碱（3）	0.20±0.04	丝氨酸	−0.15±0.01

排名基于乘以log2（t值）和log2（静脉/动脉）比值，以反映统计显著性和倍数变化。括号中的数字指的是在所示器官中表现出统计学上显著动静脉差异的代谢物数量。表中包含的所有动静脉差异在统计上都是显著的（FDR<0.05）。来自 Table 2 of Jang C, Hui S, Zeng X, et al. Metabolite exchange between mammalian organs quantified in pigs. Cell Metab 2019:1-13. https://doi.org/10.1016/j.cmet.2019.06.002.

目前对各个器官免疫细胞群落的研究还很少。我们希望在不久的将来，能够同时研究免疫细胞和微生物在身体的各部位的分布和作用。细菌被免疫细胞吞噬后，抗原表位可以由黑色素瘤的 MHC Ⅰ 递呈[97]（图4.7）。

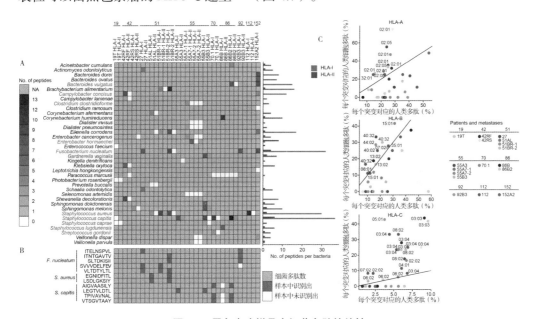

图 4.7　黑色素瘤样品中细菌多肽的特性

质谱法分析了蛋白质组，并与经过16S rRNA基因扩增子测序、过滤和验证步骤鉴定的细菌蛋白质组进行了比对。A. 在每个患者样本中，用蓝色表示（左）在HLA-Ⅰ和HLA-Ⅱ上出现的细菌多肽数量（患者数量表示在顶部），白色表示样本中没有检测到多肽，灰色表示在这种转移灶中没有检测到细菌（NA代表不适用）。每个细菌的HLA-Ⅰ和HLA-Ⅱ多肽的总数在右边的条形图中显示。用红色标记的物种名称是已知的细胞内细菌（补充表6）[97]。B. 在同一患者或不同患者的少数转移瘤中检测到的细菌肽。样本中确定的肽用绿色表示，白色表示样本中未确定的肽（尽管转移具有产生这种肽所需的HLA等位基因和细菌种类），灰色表示样品缺少HLA等位基因和细菌产生的肽。C. 对于每个转移灶，显示与患者HLA-A（左）、HLA-B（中）或HLA-C（右）等位基因相匹配的细菌和人类多肽的百分比。具有最佳等级结合预测百分比的等位基因由NetMHCpan分配给每个肽；完整的等位基因列表在扩展数据中显示。来源：Nature, 2021, 592:138-143. https://doi.org/10.1038/s41586-021-03368-8.

对于癌症和与衰老相关的疾病，不同组织的DNA突变模式也暗示了特定微生物的存在[98]。

> **思考题 4.4**
>
> 在肾结石患者中，能否将其粪便、尿液和口腔微生物与结石类型匹配，如草酸钙（二水合物、一水合物，以及更复杂类型）、磷酸钙、鸟粪石或尿酸结石[99]？我们应该考虑哪些生活方式因素？我们能给患者哪些有用的建议？（相关内容见第7章及第8章）。

4.7 总结

本章的关注点在于确定特定微生物组分的来源，可以来自环境，可以来自家庭成员，也可以来自本人的其他部位。第3章介绍的知识可帮助我们获取各种类型的样本。很多细菌可以进入淋巴结或循环系统。在其他部位出现严重症状之前，真菌可以潜伏在肠道或其他黏膜。对于类风湿性关节炎患者，滑膜液中检测到的微生物在口腔或粪便微生物组也有报道，多种口腔微生物在动脉粥样硬化斑块中有检测到。代谢物的流动及组织内的免疫细胞也可能为寻找病原菌提供线索，方便我们更好地理解和治疗疾病。

原著参考文献

[1] Afshinnekoo E, Meydan C, Chowdhury S, Jaroudi D, Boyer C, Bernstein N, et al. Geospatial resolution of human and bacterial diversity with city-scale metagenomics. Cell Syst 2015;1:1–15. https://doi.org/10.1016/j.cels.2015.01.001.

[2] Ugboko HU, Nwinyi OC, Oranusi SU, Oyewale JO. Childhood diarrhoeal diseases in developing countries. Heliyon 2020;6. https://doi.org/10.1016/j.heliyon.2020.e03690, e03690.

[3] Moossavi S, Sepehri S, Robertson B, Bode L, Goruk S, Field CJ, et al. Composition and variation of the human milk microbiota are influenced by maternal and early-life factors. Cell Host Microbe 2019;25:324–335.e4. https://doi.org/10.1016/j.chom.2019.01.011.

[4] Boix-Amorós A, Collado MC, Mira A. Relationship between milk microbiota, bacterial load, macronutrients, and human cells during lactation. Front Microbiol 2016;7. https://doi.org/10.3389/fmicb.2016.00492.

[5] Boix-Amorós A, Martinez-Costa C, Querol A, Collado MC, Mira A. Multiple approaches detect the presence of fungi in human breastmilk samples from healthy mothers. Sci Rep 2017;7:13016.

　　　　https://doi.org/10.1038/s41598-017-13270-x.

［6］Ferretti P, Pasolli E, Tett A, Asnicar F, Gorfer V, Fedi S, et al. Mother-to-infant microbial transmission from different body sites shapes the developing infant gut microbiome. Cell Host Microbe 2018;24:133–145.e5. https://doi.org/10.1016/j.chom.2018.06.005.

［7］Pannaraj PS, Li F, Cerini C, Bender JM, Yang S, Rollie A, et al. Association between breast milk bacterial communities and establishment and development of the infant gut microbiome. JAMA Pediatr 2017;171:647. https://doi.org/10.1001/jamapediatrics.2017.0378.

［8］Atarashi K, Suda W, Luo C, Kawaguchi T, Motoo I, Narushima S, et al. Ectopic colonization of oral bacteria in the intestine drives T H 1 cell induction and inflammation. Science 2017;358:359–365. https://doi.org/10.1126/science.aan4526.

［9］Liu X, Tang S, Zhong H, Tong X, Jie Z, Ding Q, et al. A genome-wide association study for gut metagenome in Chinese adults illuminates complex diseases. Cell Discov 2021;7:9. https://doi.org/10.1038/s41421-020-00239-w.

［10］Rühlemann MC, Hermes BM, Bang C, Doms S, Moitinho-Silva L, Thingholm LB, et al. Genome-wide association study in 8956 German individuals identifies influence of ABO histo-blood groups on gut microbiome. Nat Genet 2021. https://doi.org/10.1038/s41588-020-00747-1.

［11］Poore GD, Kopylova E, Zhu Q, Carpenter C, Fraraccio S, Wandro S, et al. Microbiome analyses of blood and tissues suggest cancer diagnostic approach. Nature 2020;579:567–574. https://doi.org/10.1038/s41586-020-2095-1.

［12］Anhê FF, Jensen BAH, Varin TV, Servant F, Van Blerk S, Richard D, et al. Type 2 diabetes influences bacterial tissue compartmentalisation in human obesity. Nat Metab 2020;2:233–242. https://doi.org/10.1038/s42255-020-0178-9.

［13］Li B, He Y, Ma J, Huang P, Du J, Cao L, et al. Mild cognitive impairment has similar alterations as Alzheimer's disease in gut microbiota. Alzheimers Dement 2019;1–10. https://doi.org/10.1016/j.jalz.2019.07.002.

［14］Ercoli G, Fernandes VE, Chung WY, Wanford JJ, Thomson S, Bayliss CD, et al. Intracellular replication of Streptococcus pneumoniae inside splenic macrophages serves as a reservoir for septicaemia. Nat Microbiol 2018;1. https://doi.org/10.1038/s41564-018-0147-1.

［15］Siggins MK, Lynskey NN, Lamb LE, Johnson LA, Huse KK, Pearson M, et al. Extracellular bacterial lymphatic metastasis drives Streptococcus pyogenes systemic infection. Nat Commun 2020;11:1–12. https://doi.org/10.1038/s41467-020-18454-0.

［16］Ubeda C, Taur Y, Jenq RR, Equinda MJ, Son T, Samstein M, Viale A, Socci ND, van den Brink MRM, Kamboj M, Pamer EG. Vancomycin-resistant Enterococcus domination of intestinal microbiota is enabled by antibiotic treatment in mice and precedes bloodstream invasion in humans. J Clin Invest 2010. https://doi.org/10.1172/JCI43918.

［17］Braniste V, Al-Asmakh M, Kowal C, Anuar F, Abbaspour A, Tóth M, et al. The gut microbiota influences blood-brain barrier permeability in mice. Sci Transl Med 2014;6. https://doi.org/10.1126/scitranslmed.3009759, 263ra158.

［18］Sohrabji F. Guarding the blood-brain barrier: a role for estrogen in the etiology of neurodegenerative disease. Gene Expr 2007;13:311–319. https://doi.org/10.3727/000000006781510723.

［19］Zhang X, Zhang D, Jia H, Feng Q, Wang D, Di Liang D, et al. The oral and gut microbiomes are perturbed in rheumatoid arthritis and partly normalized after treatment. Nat Med 2015;21:895–905. https://doi.org/10.1038/nm.3914.

[20] Wang J, Jia H. Metagenome-wide association studies: fine-mining the microbiome. Nat Rev Microbiol 2016;14:508–522. https://doi.org/10.1038/nrmicro.2016.83.

[21] Bostanci N, Krog MC, Hugerth LW, Bashir Z, Fransson E, Boulund F, et al. Dysbiosis of the human oral microbiome during the menstrual cycle and vulnerability to the external exposures of smoking and dietary sugar. Front Cell Infect Microbiol 2021;11. https://doi.org/10.3389/fcimb.2021.625229.

[22] Scher JU, Joshua V, Artacho A, Abdollahi-Roodsaz S, Öckinger J, Kullberg S, et al. The lung microbiota in early rheumatoid arthritis and autoimmunity. Microbiome 2016;4:60. https://doi.org/10.1186/s40168-016-0206-x.

[23] Martinez-Martinez RE, Abud-Mendoza C, Patiño-Marin N, Rizo-Rodríguez JC, Little JW, Loyola-Rodríguez JP. Detection of periodontal bacterial DNA in serum and synovial fluid in refractory rheumatoid arthritis patients. J Clin Periodontol 2009;36:1004–1010. https://doi.org/10.1111/j.1600-051X.2009.01496.x.

[24] Zhao Y, Chen B, Li S, Yang L, Zhu D, Wang Y, et al. Detection and characterization of bacterial nucleic acids in culture-negative synovial tissue and fluid samples from rheumatoid arthritis or osteoarthritis patients. Sci Rep 2018;8:14305. https://doi.org/10.1038/s41598-018-32675-w.

[25] Scher JU, Sczesnak A, Longman RS, Segata N, Ubeda C, Bielski C, et al. Expansion of intestinal Prevotella copri correlates with enhanced susceptibility to arthritis. Elife 2013;2:e01202. https://doi.org/10.7554/eLife.01202.

[26] Konig MF, Abusleme L, Reinholdt J, Palmer RJ, Teles RP, Sampson K, et al. Aggregatibacter actinomycetemcomitans-induced hypercitrullination links periodontal infection to autoimmunity in rheumatoid arthritis. Sci Transl Med 2016;8:369ra176. https://doi.org/10.1126/scitranslmed.aaj1921.

[27] Ito HO, Shuto T, Takada H, Koga T, Aida Y, Hirata M, et al. Lipopolysaccharides from Porphyromonas gingivalis, Prevotella intermedia and Actinobacillus actinomycetemcomitans promote osteoclastic differentiation in vitro. Arch Oral Biol 1996;41:439–444. https://doi.org/10.1016/0003-9969(96)00002-7.

[28] Xiao E, Mattos M, Vieira GHA, Chen S, Corrêa JD, Wu Y, et al. Diabetes enhances IL-17 expression and alters the oral microbiome to increase its pathogenicity. Cell Host Microbe 2017;22:120–128.e4. https://doi.org/10.1016/j.chom.2017.06.014.

[29] Williams DW, Greenwell-Wild T, Brenchley L, Dutzan N, Overmiller A, Sawaya AP, et al. Human oral mucosa cell atlas reveals a stromal-neutrophil axis regulating tissue immunity. Cell 2021. https://doi.org/10.1016/j.cell.2021.05.013.

[30] Qin N, Yang F, Li A, Prifti E, Chen Y, Shao L, et al. Alterations of the human gut microbiome in liver cirrhosis. Nature 2014;513:59–64. https://doi.org/10.1038/nature13568.

[31] Bajaj JS, Betrapally NS, Hylemon PB, Heuman DM, Daita K, White MB, et al. Salivary microbiota reflects changes in gut microbiota in cirrhosis with hepatic encephalopathy. Hepatology 2015;62:1260–1271. https://doi.org/10.1002/hep.27819.

[32] Bajaj JS, Acharya C, Fagan A, White MB, Gavis E, Heuman DM, et al. Proton pump inhibitor

initiation and withdrawal affects gut microbiota and readmission risk in cirrhosis. Am J Gastroenterol 2018;113:1177–1186. https://doi.org/10.1038/s41395-018-0085-9.

[33] Bajaj JS, Cox IJ, Betrapally NS, Heuman DM, Schubert ML, Ratneswaran M, et al. Systems biology analysis of omeprazole therapy in cirrhosis demonstrates significant shifts in gut microbiota composition and function. Am J Physiol Gastrointest Liver Physiol 2014;307:G951–7. https://doi.org/10.1152/ajpgi.00268.2014.

[34] Bajaj JS, Matin P, White MB, Fagan A, Golob Deeb J, Acharya C, et al. Periodontal therapy favorably modulates the oral-gut-hepatic axis in cirrhosis. Am J Physiol Liver Physiol 2018. https://doi.org/10.1152/ajpgi.00230.2018. ajpgi.00230.2018.

[35] Chhibber-Goel J, Singhal V, Bhowmik D, Vivek R, Parakh N, Bhargava B, et al. Linkages between oral commensal bacteria and atherosclerotic plaques in coronary artery disease patients. NPJ Biofilms Microbiomes 2016;2:7. https://doi.org/10.1038/s41522-016-0009-7.

[36] Rath SK, Mukherjee M, Kaushik R, Sen S, Kumar M. Periodontal pathogens in atheromatous plaque. Indian J Pathol Microbiol 2014;57:259–264. https://doi.org/10.4103/0377-4929.134704.

[37] Figuero E, Sánchez-Beltrán M, Cuesta-Frechoso S, Tejerina JM, del Castro JA, Gutiérrez JM, et al. Detection of periodontal bacteria in atheromatous plaque by nested polymerase chain reaction. J Periodontol 2011;82:1469–1477. https://doi.org/10.1902/jop.2011.100719.

[38] Ao M, Miyauchi M, Inubushi T, Kitagawa M, Furusho H, Ando T, et al. Infection with Porphyromonas gingivalis exacerbates endothelial injury in obese mice. PLoS One 2014;9. https://doi.org/10.1371/journal.pone.0110519, e110519.

[39] Bartova J, Sommerova P, Lyuya-Mi Y, Mysak J, Prochazkova J, Duskova J, et al. Periodontitis as a risk factor of atherosclerosis. J Immunol Res 2014;2014:1–9. https://doi.org/10.1155/2014/636893.

[40] Ford PJ, Gemmell E, Chan A, Carter CL, Walker PJ, Bird PS, et al. Inflammation, heat shock proteins and periodontal pathogens in atherosclerosis: an immunohistologic study. Oral Microbiol Immunol 2006;21:206–211. https://doi.org/10.1111/j.1399-302X.2006.00276.x.

[41] Taylor-Robinson D, Aduse-Opoku J, Sayed P, Slaney JM, Thomas BJ, Curtis MA. Oro-dental bacteria in various atherosclerotic arteries. Eur J Clin Microbiol Infect Dis 2002;21:755–757. https://doi.org/10.1007/s10096-002-0810-5.

[42] Haraszthy VI, Zambon JJ, Trevisan M, Zeid M, Genco RJ. Identification of periodontal pathogens in atheromatous plaques. J Periodontol 2000;71:1554–1560. https://doi.org/10.1902/jop.2000.71.10.1554.

[43] Nakano K, Inaba H, Nomura R, Nemoto H, Takeda M, Yoshioka H, et al. Detection of cariogenic Streptococcus mutans in extirpated heart valve and atheromatous plaque specimens. J Clin Microbiol 2006;44:3313–3317. https://doi.org/10.1128/JCM.00377-06.

[44] Teles R, Wang CY. Mechanisms involved in the association between periodontal diseases and cardiovascular disease. Oral Dis 2011;17:450–461. https://doi.org/10.1111/j.1601-0825.2010.01784.x.

[45] Gaetti-Jardim E, Marcelino SL, Feitosa ACR, Romito GA, Avila-Campos MJ. Quantitative detection of periodontopathic bacteria in atherosclerotic plaques from coronary arteries. J Med Microbiol 2009;58:1568–1575. https://doi.org/10.1099/jmm.0.013383-0.

[46] Mahendra J, Mahendra L, Kurian V, Jaishankar K, Mythilli R. 16S rRNA-based detection of oral pathogens in coronary atherosclerotic plaque. Indian J Dent Res 2010;21:248. https://doi.org/10.4103/0970-9290.66649.

[47] Ishihara K, Nabuchi A, Ito R, Miyachi K, Kuramitsu HK, Okuda K. Correlation between detection rates of periodontopathic bacterial DNA in coronary stenotic artery plaque [corrected] and in dental plaque samples. J Clin Microbiol 2004;42:1313–1315. https://doi.org/10.1128/JCM.42.3.1313-1315.2004.

[48] Kuo C, Campbell LA. Is infection with Chlamydia pneumoniae a causative agent in atherosclerosis? Mol Med Today 1998;4:426–430. https://doi.org/10.1016/s1357-4310(98)01351-3.

[49] Ott SJ, El Mokhtari NE, Musfeldt M, Hellmig S, Freitag S, Rehman A, et al. Detection of diverse bacterial signatures in atherosclerotic lesions of patients with coronary heart disease. Circulation 2006;113:929–937. https://doi.org/10.1161/CIRCULATIONAHA.105.579979.

[50] Schumacher A, Seljeflot I, Lerkerød AB, Sommervoll L, Otterstad JE, Arnesen H. Does infection with Chlamydia pneumoniae and/or Helicobacter pylori increase the expression of endothelial cell adhesion molecules in humans? Clin Microbiol Infect 2002;8:654–661. https://doi.org/10.1046/j.1469-0691.2002.00439.x.

[51] Farsak B, Yildirir A, Akyön Y, Pinar A, Oç M, Böke E, et al. Detection of Chlamydia pneumoniae and Helicobacter pylori DNA in human atherosclerotic plaques by PCR. J Clin Microbiol 2000;38:4408–4411. https://doi.org/10.1128/JCM.38.12.4408-4411.2000.

[52] Cochrane M, Pospischil A, Walker P, Gibbs H, Timms P. Distribution of Chlamydia pneumoniae DNA in atherosclerotic carotid arteries: significance for sampling procedures. J Clin Microbiol 2003;41:1454–1457. https://doi.org/10.1128/JCM.41.4.1454-1457.2003.

[53] Jackson LA, Campbell LA, Kuo CC, Rodriguez DI, Lee A, Grayston JT. Isolation of Chlamydia pneumoniae from a carotid endarterectomy specimen. J Infect Dis 1997;176:292–295. https://doi.org/10.1086/517270.

[54] Muhlestein JB, Hammond EH, Carlquist JF, Radicke E, Thomson MJ, Karagounis LA, et al. Increased incidence of Chlamydia species within the coronary arteries of patients with symptomatic atherosclerotic versus other forms of cardiovascular disease. J Am Coll Cardiol 1996;27:1555–1561. https://doi.org/10.1016/0735-1097(96)00055-1.

[55] Dobrilovic N, Vadlamani L, Meyer M, Wright CB. Chlamydia pneumoniae in atherosclerotic carotid artery plaques: high prevalence among heavy smokers. Am Surg 2001;67:589–593.

[56] Oshima T, Ozono R, Yano Y, Oishi Y, Teragawa H, Higashi Y, et al. Association of Helicobacter pylori infection with systemic inflammation and endothelial dysfunction in healthy male subjects. J Am Coll Cardiol 2005;45:1219–1222. https://doi.org/10.1016/j.jacc.2005.01.019.

[57] Mahendra J, Mahendra L, Nagarajan A, Mathew K. Prevalence of eight putative periodontal pathogens in atherosclerotic plaque of coronary artery disease patients and comparing them with noncardiac subjects: a case-control study. Indian J Dent Res 2015;26:189. https://doi.org/10.4103/0970-9290.159164.

[58] Okuda K, Kato T, Ishihara K. Involvement of periodontopathic biofilm in vascular diseases. Oral Dis 2004;10:5–12. https://doi.org/10.1046/j.1354-523x.2003.00979.x.

[59] Mahendra J, Mahendra L, Kurian VM, Jaishankar K, Mythilli R. Prevalence of periodontal

pathogens in coronary atherosclerotic plaque of patients undergoing coronary artery bypass graft surgery. J Oral Maxillofac Surg 2009;8:108–113. https://doi.org/10.1007/s12663-009-0028-5.

[60] Serra e Silva Filho W, Casarin RCV, Nicolela EL, Passos HM, Sallum AW, Gonçalves RB. Microbial diversity similarities in periodontal pockets and atheromatous plaques of cardiovascular disease patients. PLoS One 2014;9. https://doi.org/10.1371/journal.pone.0109761, e109761.

[61] Rafferty B, Dolgilevich S, Kalachikov S, Morozova I, Ju J, Whittier S, et al. Cultivation of Enterobacter hormaechei from human atherosclerotic tissue. J Atheroscler Thromb 2011;18:72–81. https://doi.org/10.5551/jat.5207.

[62] Latronico M, Segantini A, Cavallini F, Mascolo A, Garbarino F, Bondanza S, et al. Periodontal disease and coronary heart disease: an epidemiological and microbiological study. New Microbiol 2007;30:221–228.

[63] Moore C, Addison D, Wilson JM, Zeluff B. First case of Fusobacterium necrophorum endocarditis to have presented after the 2nd decade of life. Tex Heart Inst J 2013;40:449–452.

[64] Samant JS, Peacock JE. Fusobacterium necrophorum endocarditis case report and review of the literature. Diagn Microbiol Infect Dis 2011;69:192–195. https://doi.org/10.1016/j.diagmicrobio.2010.09.014.

[65] Stuart G, Wren C. Endocarditis with acute mitral regurgitation caused by Fusobacterium necrophorum. Pediatr Cardiol 1992;13:230–232. https://doi.org/10.1007/BF00838782.

[66] Ameriso SF, Fridman EA, Leiguarda RC, Sevlever GE. Detection of Helicobacter pylori in human carotid atherosclerotic plaques. Stroke 2001;32:385–391. https://doi.org/10.1161/01.str.32.2.385.

[67] Martínez Torres A, Martínez GM. Helicobacter pylori: ¿un nuevo factor de riesgo cardiovascular? Rev Española Cardiol 2002;55:652–656. https://doi.org/10.1016/S0300-8932(02)76673-6.

[68] Momiyama Y, Ohmori R, Taniguchi H, Nakamura H, Ohsuzu F. Association of Mycoplasma pneumoniae infection with coronary artery disease and its interaction with chlamydial infection. Atherosclerosis 2004;176:139–144. https://doi.org/10.1016/j.atherosclerosis.2004.04.019.

[69] Higuchi-dos-Santos MH, Pierri H, de Higuchi ML, Nussbacher A, Palomino S, Sambiase NV, et al. Chlamydia pneumoniae e Mycoplasma pneumoniae nos nódulos de calcificação da estenose da valva aórtica. Arq Bras Cardiol 2005;84. https://doi.org/10.1590/S0066-782X2005000600002.

[70] Kong HJ, Choi KK, Park SH, Lee JY, Choi GW. Gene expression of human coronary artery endothelial cells in response to Porphyromonas endodontalis invasion. J Korean Acad Conserv Dent 2009;34:537. https://doi.org/10.5395/JKACD.2009.34.6.537.

[71] Toyofuku T, Inoue Y, Kurihara N, Kudo T, Jibiki M, Sugano N, et al. Differential detection rate of periodontopathic bacteria in atherosclerosis. Surg Today 2011;41:1395–1400. https://doi.org/10.1007/s00595-010-4496-5.

[72] Curran SA, Hollan I, Erridge C, Lappin DF, Murray CA, Sturfelt G, et al. Bacteria in the adventitia of cardiovascular disease patients with and without rheumatoid arthritis. PLoS One 2014;9. https://doi.org/10.1371/journal.pone.0098627, e98627.

[73] Igari K, Kudo T, Toyofuku T, Inoue Y, Iwai T. Association between periodontitis and the development of systemic diseases. Oral Biol Dent 2014;2:4. https://doi.org/10.7243/2053-5775-

[74] Hans M, Madaan HV. Epithelial antimicrobial peptides: guardian of the oral cavity. Int J Pept 2014;2014:370297. https://doi.org/10.1155/2014/370297.

[75] Koren O, Spor A, Felin J, Fak F, Stombaugh J, Tremaroli V, et al. Human oral, gut, and plaque microbiota in patients with atherosclerosis. Proc Natl Acad Sci U S A 2011;108:4592–4598. https://doi.org/10.1073/pnas.1011383107.

[76] Ismail F, Baetzner C, Heuer W, Stumpp N, Eberhard J, Winkel A, et al. 16S rDNA-based metagenomic analysis of human oral plaque microbiota in patients with atherosclerosis and healthy controls. Indian J Med Microbiol 2012;30:462–466. https://doi.org/10.4103/0255-0857.103771.

[77] Mark Welch JL, Ramírez-Puebla ST, Borisy GG. Oral microbiome geography: micron-scale habitat and niche. Cell Host Microbe 2020;28:160–168. https://doi.org/10.1016/j.chom.2020.07.009.

[78] Zhu J. Over 50000 metagenomically assembled draft genomes for the human oral microbiome reveal new taxa and a male-specific bacterium; 2021. p. 2790.

[79] Jie Z, Xia H, Zhong SL, Feng Q, Li S, Liang S, et al. The gut microbiome in atherosclerotic cardiovascular disease. Nat Commun 2017;8:845. https://doi.org/10.1038/s41467-017-00900-1.

[80] Zhu W, Gregory JC, Org E, Buffa JA, Gupta N, Wang Z, et al. Gut microbial metabolite TMAO enhances platelet hyperreactivity and thrombosis risk. Cell 2016;165:111–124. https://doi.org/10.1016/j.cell.2016.02.011.

[81] Fardini Y, Chung P, Dumm R, Joshi N, Han YW. Transmission of diverse oral bacteria to murine placenta: evidence for the oral microbiome as a potential source of intrauterine infection. Infect Immun 2010;78:1789–1796. https://doi.org/10.1128/IAI.01395-09.

[82] Chen X, Li P, Liu M, Zheng H, He Y, Chen M-XX, et al. Gut dysbiosis induces the development of pre-eclampsia through bacterial translocation. Gut 2020;69:513–522. https://doi.org/10.1136/gutjnl-2019-319101.

[83] Ravel J, Gajer P, Abdo Z, Schneider GM, Koenig SSK, Mcculle SL, et al. Vaginal microbiome of reproductive-age women. Proc Natl Acad Sci U S A 2010;108:4680–4687. http://www.pnas.org/cgi/doi/10.1073/pnas.1002611107.

[84] Fredricks DN, Fiedler TL, Marrazzo JM. Molecular identification of bacteria associated with bacterial vaginosis. N Engl J Med 2005;353:1899–1911. https://doi.org/10.1056/NEJMoa043802.

[85] Jie Z, Chen C, Hao L, Li F, Song L, Zhang X, et al. Life history recorded in the vagino-cervical microbiome along with multi-omics. Genomics Proteomics Bioinformatics 2021. https://doi.org/10.1016/j.gpb.2021.01.005.

[86] Abdelmaksoud AA, Girerd PH, Garcia EM, Brooks JP, Leftwich LM, Sheth NU, et al. Association between statin use, the vaginal microbiome, and Gardnerella vaginalis vaginolysin-mediated cytotoxicity. PLoS One 2017;12. https://doi.org/10.1371/journal.pone.0183765, e0183765.

[87] Qin J, Li R, Raes J, Arumugam M, Burgdorf KSS, Manichanh C, et al. A human gut microbial gene catalogue established by metagenomic sequencing. Nature 2010;464:59–65. https://doi.org/10.1038/nature08821.

［88］The Human Microbiome Project Consortium. Structure, function and diversity of the healthy human microbiome. Nature 2012;486:207–214. https://doi.org/10.1038/nature11234.

［89］Byrd AL, Belkaid Y, Segre JA. The human skin microbiome. Nat Rev Microbiol 2018;16:143–155. https://doi.org/10.1038/nrmicro.2017.157.

［90］Rao C, Coyte KZ, Bainter W, Geha RS, Martin CR, Rakoff-Nahoum S. Multi-kingdom ecological drivers of microbiota assembly in preterm infants. Nature 2021. https://doi.org/10.1038/s41586-021-03241-8.

［91］Buffie CG, Bucci V, Stein RR, McKenney PT, Ling L, Gobourne A, et al. Precision microbiome reconstitution restores bile acid mediated resistance to Clostridium difficile. Nature 2014;517:205–208. https://doi.org/10.1038/nature13828.

［92］Hryckowian AJ, Van Treuren W, Smits SA, Davis NM, Gardner JO, Bouley DM, et al. Microbiota-accessible carbohydrates suppress Clostridium difficile infection in a murine model. Nat Microbiol 2018;3:662–669. https://doi.org/10.1038/s41564-018-0150-6.

［93］Zuo T, Wong SH, Lam LYK, Lui R, Cheung K, Tang W, et al. Bacteriophage transfer during fecal microbiota transplantation is associated with treatment response in Clostridium difficile infection. Gut 2017. https://doi.org/10.1136/gutjnl-2017-313952.

［94］Jang C, Hui S, Zeng X, Cowan AJ, Wang L, Chen L, et al. Metabolite exchange between mammalian organs quantified in pigs. Cell Metab 2019;1–13. https://doi.org/10.1016/j.cmet.2019.06.002.

［95］Rappez L, Stadler M, Triana S, Phapale P, Heikenwalder M, Alexandrov T. Spatial single-cell profiling of intracellular metabolomes in situ. BioRxiv 2019. https://doi.org/10.1101/510222.

［96］Liu X, Tong X, Zou Y, Lin X, Zhao H, Tian L, et al. Inter-determination of blood metabolite levels and gut microbiome supported by Mendelian randomization. BioRxiv 2020. https://doi.org/10.1101/2020.06.30.181438. 2020.06.30.

［97］Kalaora S, Nagler A, Nejman D, Alon M, Barbolin C, Barnea E, et al. Identification of bacteria-derived HLA-bound peptides in melanoma. Nature 2021;592:138–143. https://doi.org/10.1038/s41586-021-03368-8.

［98］Barrett M, Hand CK, Shanahan F, Murphy T, O'Toole PW. Mutagenesis by microbe: the role of the microbiota in shaping the cancer genome. Trends Cancer 2020;6:277–287. https://doi.org/10.1016/j.trecan.2020.01.019.

［99］Sivaguru M, Saw JJ, Wilson EM, Lieske JC, Krambeck AE, Williams JC, et al. Human kidney stones: a natural record of universal biomineralization. Nat Rev Urol 2021;2021:1–29. https://doi.org/10.1038/s41585-021-00469-x.

第 5 章

演化中的微生物分类

摘　要：本章介绍了微生物分类学在人类宏基因组学研究中的现状，从培养的菌株和每个身体部位的宏基因组装结果中得出参考基因组。在第 7、第 8 章中讨论的应用，需要满足两个先决条件：一个是全面的数据库，一个是足够快的算法。尽管存在潜在的可移动的遗传因子，微生物物种分类仍然是稳定的实体。低于物种水平的变异，通常被称为菌株，可以通过宏基因组测序或培养分离，在同一个人身上进行长期追踪。更高的分类学分辨率导致相对丰度中更多的零值，这需要适当的统计学方法处理。随着对共生微生物进行大量体外培养和身体多位置研究，基于生物信息学和代谢建模的分类学将最终包含基于特定功能检测的传统微生物分类。

关键词：细菌基因组，宏基因组组装基因组（MAGs），培养组学，真菌分类学，细菌种类，细菌菌株，稀疏性，机器学习，全细胞模型

5.1　对于常规应用的固定参考集

为了满足临床数据分析快速需求，目前常用方法是将宏基因组测序得到的读段（reads）或者组装形成的基因序列，直接比对到一组现有的参考序列。16S rRNA 基因扩增子测序尚无确定的参考序列（图 5.1）[2,3]，因此需要将序列聚类到操作分类单位（OTUs），然后利用推断出的唯一"种子序列"与数据库中已经有的科、属或物种进行映射，而剩下不同部分无法确定的 OTUs。宏基因组鸟枪法测序不受 PCR 扩增偏向性的影响[4]，能够检测真核生物和病毒，可以对应微生物基因组的所有部位。据覃俊杰博士介绍，2010 年代初，他曾要求李胜辉展示数百个样本中，根据相对丰度的共变化而聚类的每个基因的分类信息，结果发现，同一聚类中的基因都来自同一种细菌[5]。这说明共变化规律能够反映微生物基因组中的物理连锁，这种连锁应该是在菌株水平上的，但是半数的聚类属于未知物种，在计算和实验上还有许多改进的空间。

如今，可以根据单一宏基因组样本组装得到的重叠群（contig），并根据序列组成（如四碱基频率）进行分箱。利用多个样本之间的共变化信息，可以优化重叠群的覆盖信息，并提高中、高质量基因组基因的数量。然而，由于有亲缘关系的微生物的

第 5 章 演化中的微生物分类

序列相似（例如来自同一样本的两个菌株的序列），可能导致宏基因算法在组装及分箱阶段产生混淆[6,7]。人类粪便微生物的参考序列，是由培养分离物的组装结果和宏基因组装得到的序列组成的[8]，这些序列在种或者株的水平去除冗余，用于基于基因组的研究（图 5.1）。

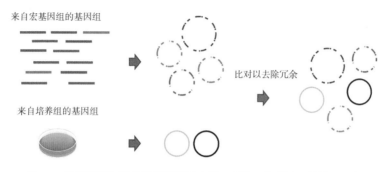

图 5.1 宏基因组和培养组的基因组来源：去除冗余后用于基因组研究

无偏地从宏基因组样本中提取微生物基因组，以获得群落的完整表征。来自高通量宏基因测序组装结果，目前尚不完善。

人体肠道微生物在门的级别主要包括厚壁菌门、拟杆菌门、变形杆菌门、放线菌门和梭杆菌门，还有纤维菌门、螺旋体门、黏胶球形菌门，以及更神秘的门类如异常球菌-栖热菌门、蓝细菌、绿弯菌门（图 5.2）。对于口腔微生物组，我们可以从候选门辐射类群（CPR）（图 5.2 和图 5.3）中看到更多的门，其中最著名的是 TM7x 属（Saccharibacteria 门，旧称 TM7），它是一种放线菌［溶牙放线菌（*Actinomyces odontolyticus*）］的表面寄生菌[11-13]。最近的一项研究认为 CPR 是通过从自由生活的基因组经由进化缩减而来的，而不是处于图 5.2[14]所示的进化树的根位置，这与它们专性共生体的生活方式和体型限制相一致[15,16]。除了产甲烷菌（第 2 章，示例 2.4），人类皮肤中还含有奇古菌，可以氧化氨产生亚硝酸盐[17]。

传统观点认为大多数微生物是"不可培养的"，要获得最佳培养条件十分困难，因此正常菌群中一些丰富的属和物种，直到近年来才有了一个或多个基因组草图[18-20]。例如，要对肺部微生物组完整实现培养组学，就需要考虑到肺部不同位置的温度、气体和 pH 梯度的变化（第 3 章，图 3.6）。真菌在培养组学和高通量分析之间的匹配道阻且长（图 5.4），更遑论病毒了。虽然宏基因组测序理论上能够检测样本中的所有微生物（第 1 章，图 1.2），但是对于一些低丰度的分类群，通过特定的培养方法更容易获得，这也有助于获取更多的基因组学和其他组学信息[10,18,21,22]。

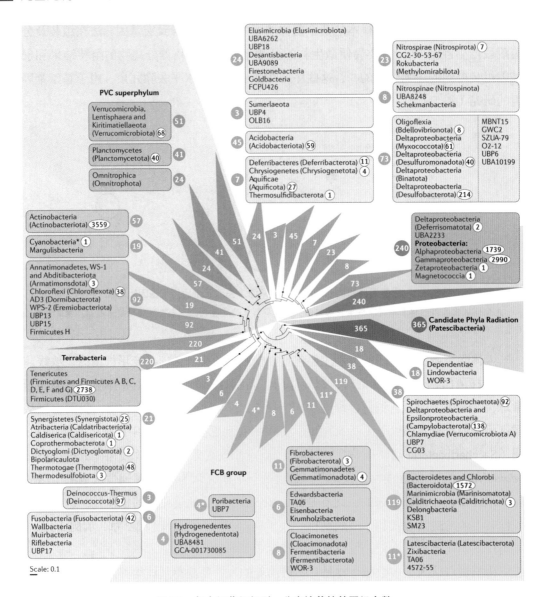

图 5.2　每个细菌门级别，分离培养的基因组个数

在不区分环境、宿主和身体部位的情况下，培养的细菌目前以拟杆菌门、变形杆菌门、厚壁菌门和放线菌门为主。细菌的系统发育树，由基因组分类数据库获得的1541个细菌基因组编码[9]构成，这些基因组来自每个物种总共15个核糖体蛋白质中至少5个比对上的序列。白色带颜色字体的数字代表每个进化枝中的独立分类单元的数量，并将对应的分类单元名称和进化枝相连。黑色白边的数字是对应描述的分离培养物中，出现在BacDive数据库[10]（2020年4月），有多少个物种被分配到对应的进化枝中的个数。没有数字的分类单元，表示没有出现在BacDive的分离培养产物中。

由于很多分离培养产物没有被正式描述或存储在分离培养产物集合中，因此它们并没有在科学文献中报道，也没有体现在图中的数字中。目前，还没有一个全面的数据库，能够记录所有已经培养过的细菌，包括那些未经正式描述或保存在分离培养集合中的细菌。本研究构建了一棵系统发育树，用于展示不同细菌的亲缘关系。该树是基于包含不同物种同源蛋白的数据集生成的，这些同源蛋白首先用MAFFT（L-INS-i）进行序列比对，然后将每个蛋白的比对结果拼接起来，使同一物种的蛋白序列合并为一个完整的序列。为了提高比对的质量，采用trimAI技术去除连接序列中保守性较差的位点，参数为＞0.5。然后，使用IQ-TREE中的LG+C60+f+r10

模型，对去除后的比对结果进行1000次超快引导复制，以进行系统发育分析，从而构建系统发育数。支撑值≥95%的枝条用黑点标记。

需要注意的是，由于用于推断这种系统发育的蛋白质数据集有限，在某些情况下，某些物种或群体之间的深层关系可能无法反映更广泛接受的基于更深入和更好支持的分析得出的发育树。特别是，异常球菌-栖热菌门和衣原体分别与地球菌属的其他谱系和PVC超门不同类群。虽然已经有大量蓝细菌血统的培养代表，但在BacDive中的代表性却特别低。这与大多数细菌不同，由于历史原因，蓝细菌大多使用植物学编码进行分类（藻类、真菌和植物的国际命名法规）。因此，蓝细菌缺乏明确的类型菌株，因此没有在BacDive中广泛列出，也缺乏现有蓝藻培养物的全面数据库。来源：Nat Rev Microbiol, 2021, 19:225-240. https://doi.org/10.1038/s41579-020-00458-8.

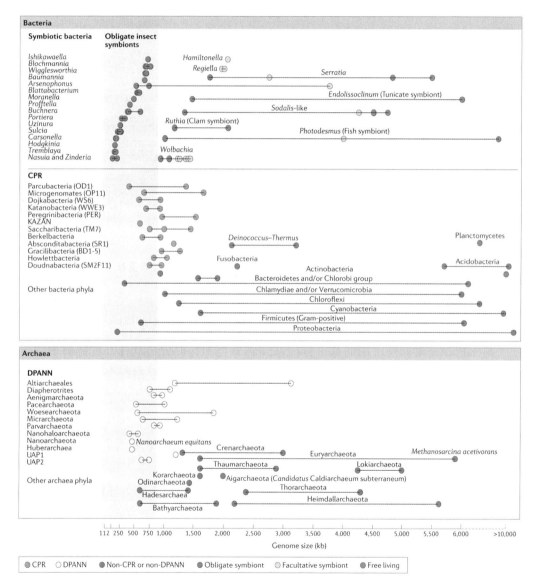

图 5.3 对细菌（CPR）和古菌（DPANN）的基因组大小，与已知的共生细菌和古菌的对比

上图显示了被充分研究的细菌的数据，这些细菌分为专性共生（橙色点）、兼性共生（绿色点）和自由生活的细菌（灰色点）。中间的子图显示了CPR信息（紫色点），底部图提供DPANN（黄色点）的基因组大小信

息。中间和底部的子图也显示了其他细菌和古菌的大小范围（蓝点）。CPR和DPANN基因组大小与专性共生菌的基因组大小重叠。来源：Nat Rev Microbiol, 2018, 16:629-645. https://doi.org/10.1038/s41579-018-0076-2.

图 5.4　使用组织培养和非组织培养方法分析肠道真菌生物群落时，在种水平上发现的真菌数据的共性和差异

这张维恩图强调了仅仅通过培养无关、仅通过培养有关的方法检测到的真菌种类，以及通过两种方法（都能够）检测到的种类。来源：FEMS Microbiol Rev, 2017, 41:479-511. https://doi.org/10.1093/femsre/fuw047.

MetaPhlAn 系列是常用的基于标记基因的分类学特征分析软件，它能够识别出 50 门，共 7677 个物种和更多的株系[23-25]。这些标记基因是根据每个分类单元的特异性预先设定的。MetaPlanAn2 为了构建标志基因集，分析了 300 个古菌基因组、12 926 个细菌基因组、3653 个病毒基因组以及 112 个原核生物基因组中进行采样。MethPhlAn3 则使用了总计 99.2 万的高质量基因组，更新了标志基因集，并对未知的分类单元比例进行了估计[23,26]。

我们期待在未来的宏基因组研究中，科和属级别的信息能够更加精确[6,7]，同时数据库和生信分析流程在未来不需要频繁更新。正如第 1 章所示，尼安德特人的口腔和粪便微生物，即使在 DNA 高度碎片化的情况下，也呈现出和现代人样本在种级别的相似性[27-30]。对于特定的应用（第 7、第 8 章），一个更小的参考基因数据库意味着更快的运行速度和更少的混淆结果。

第 5 章 演化中的微生物分类

思考题 5.1

以下是一个鼻拭子宏基因组的相对丰度。使用现有的多种软件，包括基于kmer（基因组中连续 k 个碱基序列）的 Kraken2+Bracken[31]，以及基于标记基因的 mOTU2[24]、MetaPhlAn2 和 MetaPhlAn3，对其进行分析。

（1）请了解不同的方法和数据库的特点，以及它们的敏感性和准确性[6,7]。你是否更倾向于属一级的结果？

（2）请对于每一种方法得到的最丰富的物种，按照相对丰度从 1 到 30 进行排名。你觉得不同的方法的结果是否更加一致呢？

随着样本数量的增加，你是否理解为什么对宏基因组数据最常用的相关系数是基于秩的？例如斯皮尔曼相关系数（如第 2 章，图 2.14），而皮尔森相关系数，包括常用的 SparCC[33]，只是线性相关。

（3）在你对正常菌群的研究中，你更关心哪个级别的分类单元？你认为哪些方面需要改进？你能猜出哪种方法更容易适应新的组装后的基因组吗？

微生物大类	微生物种	Kraken2+Bracken	mOTU2	MetaPhlAn2	MetaPhlAn3
Bacteria	*Acinetobacter baumannii*	0.157951	0	0	0
Bacteria	*Anaerococcus* species incertae sedis (uncertain placement of species) [meta mOTU v25 12712]	0	0.454865	0	0
Bacteria	*Bacilli* sp. [ref mOTU v25 00344]	0	0.239723	0	0
Bacteria	*Bacillus cereus*	0.130481	0	0	0
Bacteria	*Bacteria* sp. [ref mOTU v25 00259]	0	0.155647	0	0
Bacteria	*Bacteria* sp. [ref mOTU v25 00964]	0	0.131472	0	0
Bacteria	*Brachybacterium paraconglomeratum*	0	0.116516	0	0
Bacteria	*Corynebacterium accolens*	0	24.27697	9.31315	32.04989
Bacteria	*Corynebacterium ammoniagenes*	0.1528	0	0	0
Bacteria	*Corynebacterium aurimucosum*	0.448099	0.490033	0	0
Bacteria	*Corynebacterium camporealensis*	0.458401	0	0	0
Bacteria	*Corynebacterium casei*	0.357106	0	0	0
Bacteria	*Corynebacterium diphtheriae*	0.559695	0	0	0
Bacteria	*Corynebacterium flavescens*	0.375992	0	0	0
Bacteria	*Corynebacterium glutamicum*	0.923669	0	0	0
Bacteria	*Corynebacterium jeikeium*	0.243794	0	0	0
Bacteria	*Corynebacterium kroppenstedtii*	1.222402	1.712234	2.09802	1.23582
Bacteria	*Corynebacterium minutissimum*	0.293582	0	0	0
Bacteria	*Corynebacterium phocae*	0.1528	0	0	0
Bacteria	*Corynebacterium propinquum*	0	0.421024	2.12244	0

续表

微生物大类	微生物种	Kraken2+Bracken	mOTU2	MetaPhlAn2	MetaPhlAn3
Bacteria	*Corynebacterium pseudogenitalium*	0	0	0.71378	0
Bacteria	*Corynebacterium resistens*	0.104728	0	0	0
Bacteria	*Corynebacterium simulans*	0.882464	0	0	0
Bacteria	*Corynebacterium singulare*	0.336504	0	0	0
Bacteria	*Corynebacterium* sp. [ref mOTU v25 03067]	0	1.811982	0	0
Bacteria	*Corynebacterium* sp. [ref mOTU v25 00802]	0	0.109947	0	0
Bacteria	*Corynebacterium stationis*	0.14765	0	0	0
Bacteria	*Corynebacterium striatum*	1.857638	0	0	0
Bacteria	*Corynebacterium ureicelerivorans*	0.108162	0	0	0
Bacteria	*Cutibacterium* (formerly *Propionibacterium*) *acnes*	12.71332	9.133923	20.45757	11.96932
Viruses	*Propionibacterium phage* BruceLethal	0.157951	0	0	0
Viruses	*Propionibacterium phage* Moyashi	0.243794	0	0	0
Viruses	*Propionibacterium phage* P101A	0	0	5.99864	0
Viruses	*Propionibacterium phage* PA1-14	0.140782	0	0	0
Viruses	*Propionibacterium phage* PHL009	0.255812	0	0	0
Viruses	*Propionibacterium phage* PHL010M04	0.456684	0	0	0
Viruses	*Propionibacterium phage* PHL030	0.441232	0	0	0
Viruses	*Propionibacterium phage* PHL055	0.104728	0	0	0
Viruses	*Propionibacterium phage* PHL070	0.489304	0	0	0
Viruses	*Propionibacterium phage* PHL082	0.118463	0	0	0
Viruses	*Propionibacterium phage* PHL085	0.396594	0	0	0
Viruses	*Propionibacterium phage* PHL116	0.400028	0	0	0
Viruses	*Propionibacterium phage* PHL132	0.209456	0	0	0
Viruses	*Propionibacterium phage* PHL141	0.157951	0	0	0
Viruses	*Propionibacterium phage* PHL152	0.679875	0	0	0
Viruses	*Propionibacterium phage* PHL171	0.127047	0	0	0
Viruses	*Propionibacterium phage* QueenBey	0.427497	0	0	0
Viruses	*Propionibacterium virus* Attacne	0.260962	0	0	0
Viruses	*Propionibacterium virus* Lauchelly	0.496171	0	0	0
Viruses	*Propionibacterium virus* Ouroboros	0.454967	0	0	0
Viruses	*Propionibacterium virus* P100A	0.108162	0	0	0
Viruses	*Propionibacterium virus* PHL071N05	0.199155	0	0	0
Viruses	*Propionibacterium virus* PHL114L00	0.204306	0	0	0
Viruses	*Propionibacterium virus* Pirate	0.338221	0	0	0
Viruses	*Propionibacterium virus* Solid	0.116746	0	0	0
Viruses	*Propionibacterium virus* Stormborn	0.257528	0	0	0

续表

微生物大类	微生物种	Kraken2+Bracken	mOTU2	MetaPhlAn2	MetaPhlAn3
Bacteria	*Cutibacterium granulosum*	0.990626	0.557325	0	1.05931
Bacteria	*Erythrobacteraceae bacterium* CCH12-C2	0	0.11621	0	0
Bacteria	*Haemophilus parainfluenzae*	0.111596	0	0.16822	0
Bacteria	*Klebsiella michiganensis/oxytoca*	0	0.124238	0	0
Bacteria	*Klebsiella oxytoca*	0.121897	0	0	0
Bacteria	*Lautropia mirabilis*	0	0.118127	0.16232	0
Bacteria	*Lawsonella clevelandensis*	0.175119	6.701121	0	0
Eukaryota-Fungi	*Malassezia restricta*	0	0	0	15.47571
Eukaryota-Fungi	*Malassezia* species incertae sedis [meta mOTU v25 12989]	0	15.03185	0	0
Bacteria	*Moraxellaceae* sp. [ref mOTU v25 06002]	0	0.109049	0	0
Bacteria	*Morococcus cerebrosus*	0	0.105722	0	0
Bacteria	*Neisseria elongata*	0.582014	0.283177	0.38064	0.42907
Bacteria	*Neisseria macacae*	0	0	0.1394	0.11948
Bacteria	*Neisseria meningitidis*	0.103011	0.231828	0	0
Bacteria	*Neisseria mucosa*	0.255812	0	0	0
Bacteria	*Neisseria sicca*	0.281564	0	0.43785	0.89061
Bacteria	*Neisseria sicca/macacae*	0	0.216081	0	0
Bacteria	*Neisseria* sp. [ref mOTU v25 04798]	0	0.169086	0	0
Bacteria	*Neisseria* sp. HMSC064E01	0	0.295144	0	0
Bacteria	*Neisseria* sp. oral taxon 014	0	0.102625	0	0
Bacteria	*Neisseria unclassified*	0	0	0.60812	0
Bacteria	*Prevotella melaninogenica*	0	0.196181	0	0
Bacteria	*Pseudomonas stutzeri*	0.18027	0.133901	0	0
Bacteria	*Sphingomonadales bacterium* RIFCSPHIGHO2 01 FULL 65 20	0	0.101038	0	0
Bacteria	*Staphylococcus aureus*	0.479003	0	0	0
Bacteria	*Staphylococcus capitis*	0.20774	0	0	0
Bacteria	*Staphylococcus epidermidis*	39.33489	33.71282	55.81221	35.78187
Bacteria	*Staphylococcus hominis*	0.307317	0	0.22698	0.16856
Viruses	*Staphylococcus phage* StB27	0.259245	0	0	0
Viruses	*Staphylococcus virus* IPLAC1C	0.382859	0	0	0
Viruses	*Staphylococcus virus* SEP9	7.926725	0	0	0
Viruses	*Staphylococcus virus* Sextaec	12.46781	0	0	0
Bacteria	*Streptococcus mitis*	0.173403	0	0	0.15237
Bacteria	*Streptococcus mitis/oralis/pneumoniae*	0	0	0.25495	0
Bacteria	*Streptococcus sanguinis/cristatus*	0	0.106522	0	0
Bacteria	*Streptococcus* sp. [ref mOTU v25 00283]	0	0.412096	0	0

思考题 5.2

请从同一个人的舌苔和唾液样本进行宏基因组测序，或下载公开数据。

（1）请分别统计在每种样本种中能检得到多少物种水平（或更高一级分类单元）的检测结果？

（2）你能否组装出某些高质量基因组，唾液和舌苔的样本组装结果会有多大差异呢？

（3）在马拉松优秀跑步者的肠道中发现的乳酸代谢的韦荣氏球菌和在自身免疫性疾病如类风湿关节炎中发现的韦荣氏球菌是否有区别？

5.2　随着分类单元精度提升，数据变得稀疏

第 3 章讨论了宏基因研究中的统计学实践。我们知道，物种和菌株的信息（见示例 5.1）对于功能描述是至关重要的，但是更细的分类单元也意味着分类单元在更多样本中丰度为零。数据的稀疏性在统计学中是一个问题。现在已有不少方法学尝试，来区分采样零值（由于测序量未检出）和真实零值（该微生物不存在）[35]。在研发针对宏基因组数据统计学方法之前，我们会看到很多 P 值，会随着分类精度变高而变大，而此时有效样本数会变小。例如，在粪便微生物组和血浆代谢物的孟德尔随机化分析中，变形菌门多次出现，通常比对应的属或种的结果更"显著"[36]。在属水平之下，由于同一个测序的读段，会被比对多个参考基因组中，会严重地影响相对丰度的值[6]。因此，我们不确定是否有多个相似的物种和菌株和人类基因相关，或在进行孟德尔随机化之前，要进行更精细的物种刻画。

更高的物种分辨率，还会影响基于多样本计算的相关性分析（包括分类单元之间的相关性，以及分类单元和其他组学的特征的相关性）。一个原本丰度很高的分类单元（如思考题 5.1 中），可能会被拆分为多个更小的、相对丰度波动性更大的分类单元，从而降低了它们在斯皮尔曼相关性或其他统计指标中的显著性。

为了从宏基因组关联分析中筛选出具有生物学意义的标志物，可以采用一些适合处理稀疏数据的机器学习方法，例如随机森林和 LASSO[37]。这些方法选出的标志物更有可能在其他数据集中得到验证，而不是仅仅适用于训练数据的过拟合结果。此外，为了增加模型的稳定性和泛化能力，常常结合随机森林使用 10 折或 5 折交叉验证的方法，每次将 10% 的样本作为测试集，剩余的样本作为训练集，重复进行多次[38-40]。

最新的神经网络算法，能够更好地挖掘宏基因组数据中多层次的内在联系，以及和其他组学数据的作用关系，对关联分析、疾病诊断和预防的生物标志物发现，以及 MAG 组装和功能注释都有重要的价值。

第 5 章 演化中的微生物分类

示例 5.1 细菌的物种概念

对于病毒和细菌，物种的定义并不是个问题。达尔文在 1859 年出版的《物种起源》[43,44] 中这样写道：

"总而言之，我认为物种是相当明确的对象，它们在任何时期都没有表现出由无数不可区分的中间形态所构成的混乱……

如果我的理论是正确的，那么必然存在着无数的中间变异，它们将同一群体的所有物种紧密地联系在一起，但是自然选择的过程本身不断地趋向于……消灭亲本和中间环节……

人们会发现，我把物种这个术语看作是为了方便起见而随意标记为一组彼此十分相似的个体，它本质上与变种（variety）这个术语并没有什么不同，后者代表不那么明显和更加不稳定的个体。这可能不是一个令人振奋的前景，但是我们至少将从对物种这个术语从对未发现和不可发现的本质的徒劳探索中解脱出来。"

阿尔弗雷德·拉塞尔·华莱士，自然选择的共同发现者，后来给出的对物种的定义，更像是一个跨代保持的生态类型：

"一个物种……是一组活着的有机体，它们通过一系列独有的特征与其他所有这样的群体中分离出来，与环境的关系也不同于任何其他有机体群体，并且具有不断繁殖的能力。"[45]

在原核生物中定义物种的一个主要困难是，有性生殖不能作为标志物种之间界限的标准。有趣的是，微生物组可以通过影响昆虫的生殖隔离、配偶鉴别和混合不育/致死，从而促进物种形成[46]。

同样值得关注的是，微生物的基因组可能更具有可塑性。基因水平转移可以通过跨越分类学界限的可移动元素实现，但似乎并不常见[16,47]。核心基因序列在物种水平上确实表现出了明显的差异（图 5.5）。因此，细菌物种拥有各自的基因组特征，不会轻易地通过突变的累积或通过基因的水平转移出现在其他物种中。核心基因还决定了代谢和细胞壁性状，这些性状在传统的微生物学中是可以被检测的（如第 1 章，图 1.11）。

生态型是生态位理论的一个概念，生态位不仅限于遗传的属性，也包括群落中单细胞水平的异质基因表达[48-50]。当基因组支持时，生态型更适合在各种条件下进行功能研究，对于定义微生物种类有辅助作用。对于生活在正常菌群的物种，功能研究可能需要包括与宿主免疫系统的相互作用，例如，如果我们首先积累了实验证据，就可以从微生物基因组中预测其抗原特性。

思考题 5.3

基于 1267 份粪便样本，根据包含 9 879 896 个基因的参考基因集（在合并时，根据 95% 的同源性，删除了"冗余基因"中较短的那个）进行了分析，我们先前估计，每两个个体的肠道微生物基因有 1/3 是共享的。每个样本平均含有 762 665 个基因，任意两个样本平均共有 250 382 个基因（占 762 665 个基因的 32.8%）[41]。

（1）在分类学的层面上，以某一特定群体为重点，你目前对于两个个体之间共享的肠道微生物门、科、属、种的数量有什么想法？你能在孢子形成的细菌[42]和更依赖垂直传播的细菌（例如母亲和婴儿之间）中观察到不同模式吗（见扩散限制，第 2 章，图 2.3）？

（2）对于身体其他位置的微生物，又有什么不同的情况呢？

5.3 物种水平以下的微生物演化历程

虽然生殖隔离的概念对细菌不适用（示例 5.1），但是在基于核心基因的基因组分析可以明确地划分出物种的界限（图 5.5）。一些属经过重新命名，从一个原有的属名中分化出来，这是基于它们的基因组差异和功能特征。普雷沃菌（*Prevotella copri*）（第 2 章）是一个例子，它至少包含 4 个分支，多年来只有一个基因组草图，现在被称为普雷沃菌复合体（图 5.6，译者注：更新的属名改成了 *Segatella*，种名 *S.copri*），以表明它不是单一物种。由于 DNA 聚合酶在基因组复制过程中的错误率为 10^{-8}，且修复率低于真核生物，在 4.6 Mb[51]的基因组中，大肠埃希菌在 1.85×10^9 个核苷酸中累积约 1 个突变。脆弱拟杆菌的基因组大小为 5.2 Mb，从相同人的时序样中重复分离的物种显示出每年积累一个 SNP 突变（图 5.7、图 5.8）。这表明，如果从同种群开始，这种肠道黏膜细菌每天只复制一次 $\{1.85 \times 10^9 / [(5.2 \times 10^6) \cdot 365] = 0.97 \approx 1\}$。更大的变异可能会导致这种细菌失去对这个生态位的适应性，例如，缺乏定殖的脆弱拟杆菌（第 1 章，图 1.1C），偶尔会发生基因水平转移[52]。那些复制频率更高或者流行规模更大的菌种可以更快地积累 SNPs。对于囊性纤维性患者肺部的铜绿假单胞菌，在肺的不同区域可以形成不同的谱系。

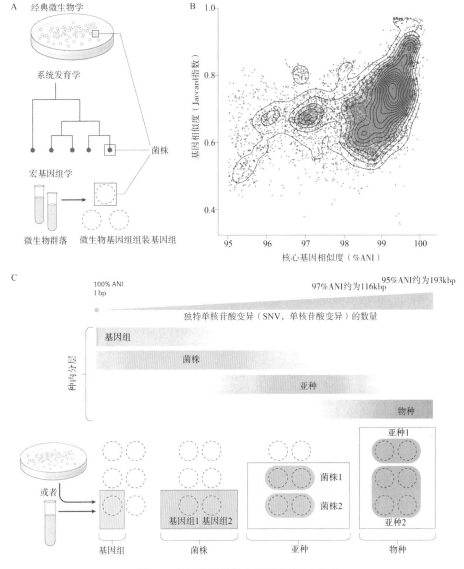

图 5.5　同一物种的微生物基因组的多样性

A. 不同的研究领域对"菌株"有不同的操作性定义：传统微生物学中的分离培养菌株，系统发生树中的叶节点菌株，和宏基因组学中的组装菌株。B. 每一个点代表一个分离基因组与其同种分离基因组的成对比较。这些数据涵盖了155种细菌，每种至少有10个已测序的分离基因组。红色地形覆盖图的不透明度反映了点的密度。图表显示了核心基因组的相似性（用平均核苷酸同一性，ANI，衡量）与基因内容相似性（用Jaccard指数衡量）的关系。核心基因序列相似性较高的基因组具有更多的共同基因（斯皮尔曼相关系数为0.57，P值小于2.2×10^{-16}）。然而，高的ANI并不保证高的基因内容相似性，许多核心基因组中的ANI超过99%的菌株对只有不到70%的共同基因。大多数物种（83%）内的ANI值超过97%，ANI低于95%的4%数据点未显示。C. 细菌种内变异的关键变异的空间分布，从全基因组的单核苷酸变异（SNV）到物种水平阈值（97%ANI）。条形图的彩色部分反映了每个分类层级的推荐使用范围，灰色部分表示常见但不具体的使用范围。从广义上讲，同种基因组97%的基因组（97%ANI）在同源位点上有相同的核苷酸，这相当于基于平均3.87 Mb的细菌基因组大小计算的116 000个SNVs的差异。底部的子图说明了这些分类单元的层次结构，一个物种可能包含多个亚种，一个亚种包含多个菌株，一个菌株包含多个（不同的）基因组。这些基因组可以通过从分离培养或通过对测序数据的组装获得。

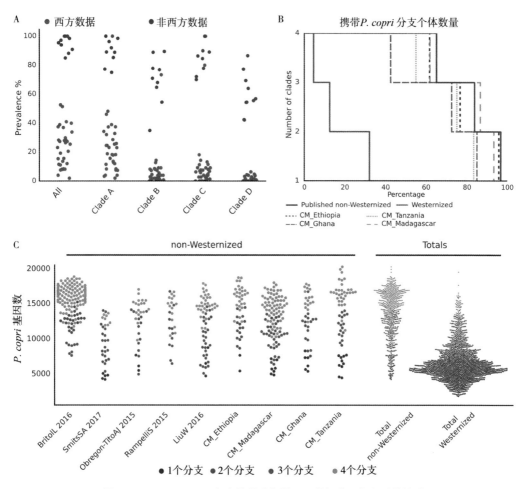

图5.6 *Prevotella copri*复合体的流行情况及其与非西化人群的关联

A.*Prevotella copri*在非西方化和西方化数据集中都普遍存在。"全部"指的是所有四个分支都存在的普遍性。B.携带多个*Prevotella copri*分支的个体百分比。C.与非西方化个体相比,西方化个体中*Prevotella copri*复合体的泛基因组大小。*Prevotella copri*复合体每个分支特有的蛋白质编码基因被定义为在给定分支的超过95%的*Prevotella copri*基因组中存在,而在所有其他分支中不存在。这给出了分支A的标记数为430,分支B的标记数为954,分支C的标记数为479,分支D的标记数为585。图片来源:https://doi.org/10.1016/j.chom.2019.08.018. https://doi.org/10.1038/s41396-020-0600-z.

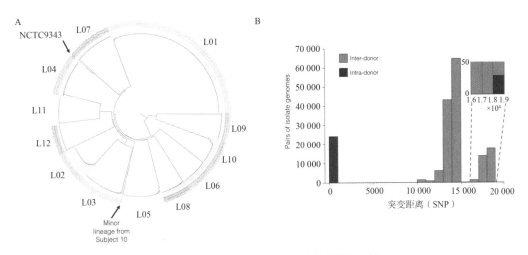

图 5.7　肠道细菌的稳定谱系和个体特异性示例

12个健康受试者的脆弱拟杆菌群落都主要来自一个谱系。其中7人的样本覆盖了两年的时间跨度。A. 系统发育重建结果表明，分离菌株按样本分成不同的类群（n=602）。分离菌株的颜色（L01～L12）表示样本的来源。左上方的箭头表示的是脆弱芽胞杆菌CCUG4856T（NCTC9343）的NCBI参考基因组。L05旁边的箭头表示来自10号被试的单个样本，它与L10中的菌株不属于同一类群。B. 来自同一受试者的分离菌株之间的单核苷酸变异一般小于100个，而来自不同受试者的分离菌株之间的单核苷酸差异一般大于10 000个。与10号受试者的分离样本相关的菌株对之间的单核苷酸变异超过18 000个，可以视为异常值。来源：Cell Host Microbe, 2019. https://doi.org/10.1016/j.chom.2019.03.007.

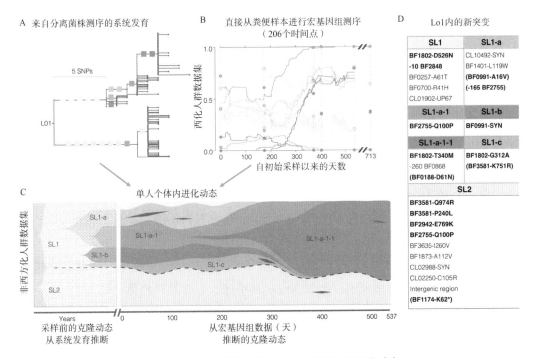

图 5.8　脆弱拟杆菌的突变频率和子细胞系的进化动态

我们分析了一个志愿者在1.5年的抽样期间的脆弱拟杆菌群落进化动态，发现它们的突变频率稳定增加，且两个子细胞系稳定共存。移动的基因元素没有显示。（A～C）我们结合了206个粪便宏基因组和187个分离

115

培养的全基因组，来推断L01聚簇内的进化动态（图5.7）。A. 彩色正方形标记了至少有4个分离株的分支上的单个SNPs。一个SNP被推断发生了两次，并用紫色在两个地点标出。B. 利用宏基因组推测标记SNP的发生频率。圈代表从孤立基因组推断出的SNP频率。C. 我们将这些数据类型结合起来，推断出抽样之前和抽样期间子细胞数量的变化轨迹。子字段用名称标记，并用A. 中的颜色标记。用一条虚线分隔了两个主要的子类，SL1和SL2。黑色矩形的点代表来自多糖利用（PULs）和细胞膜生物合成基因的瞬时SNPs。D. A到C中单核苷酸多态性（SNPs）的分类。用粗体表示了正向选择下的16个基因中的SNPs，用细体表示了这些基因中的瞬时突变。负数代表发生在基因起始位点之前的突变（上游区域）。来源：Cell Host Microbe, 2019. https://doi.org/10.1016/j.chom.2019.03.007.。

通过质粒、噬菌体或其他可移动元件的基因水平转移，更易形成在压力条件下有明显优势的功能，如抗生素抗性[54,55]。当选择压不再存在时，这种作用可能成为一种负担；当选择压存在时，丰度较高的菌种并不一定是能抵抗变化的[56]。

5.4 全细胞模型以实现从基因组到功能的完整预测

生殖支原体是拥有525个基因的小型基因组，它们的研究论文超过900篇，它也是第一个用计算机模拟细胞中每个主要过程的有机体（图5.9）[57]，其中27.5%的参数是基于生殖支原体自身的实验数据，而不是其他细菌的实验。最近的大肠埃希菌模型涵盖了1214个基因（占有良好注释的大肠埃希菌基因的43%），并从几十年来关于大肠埃希菌本身的文献中收集了超过19 000个观测得到的参数值[58]。对大多数人体共生微生物，这样详尽的研究还是很难实现的，其中很多只有一篇命名该种微生物的文献。这类模型，相比只考虑代谢流的基于约束的模型，更为确定，可能成为连通微生物基因组和表型，并指导未来的实验设计的关键[57,58]。生殖支原体研究预测了必需基因和非必需基因，并从差异中识别出了一些以前未知的冗余功能[57]。大肠埃希菌模型揭示了一些之前的研究和模拟结果都忽略的现象，例如已报道的核糖体和RNA聚合酶的数量不足以支持复制所需时间，并发现许多必需的蛋白质在细胞周期内没有被转录，甚至在某些细胞中可能缺失[58]。例如，编码4-氨基-4-脱氧氯索酸合成酶（pabA和pabB）的基因，每个细胞周期转录频率分别为0.94和0.66次，平均每代产生34个PabA蛋白和101个PabB蛋白；在没有PabAB二聚体的情况下，5,10-二亚甲基四氢叶酸（methylene-THF）随着时间消耗。

大肠埃希菌细胞中，大分子（200～300 g/L，高于图1.4中的模拟）如此密集地充满细胞，以至于其呈玻璃态。大于30 nm的分子，如多聚核糖体（每个21 nm）、大型酶复合物、质粒和噬菌体颗粒会停留在原位而不是扩散，与自由扩散的小分子[61]形成鲜明对比（图5.10）。代谢活动使玻璃态流动并增加运动[61]。哺乳动物细胞的细胞核也呈类似的玻璃态。如果全细胞模型有足够的实验数据来预测物理特性，那么它们最终将能够预测菌落中的动力学，以及与宿主上皮细胞的复杂相互作用。

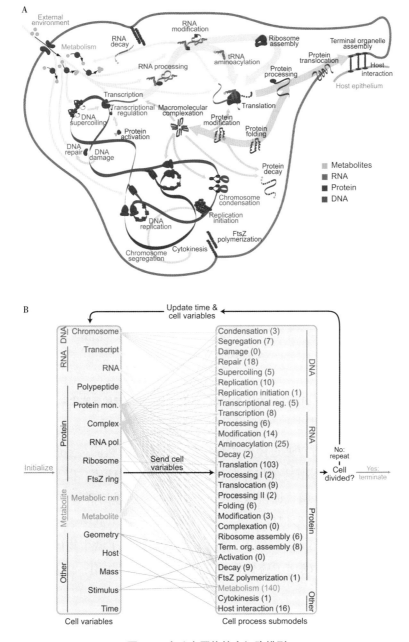

图 5.9 生殖支原体的全细胞模型

A. 生殖支原体全细胞模型集成了 28 个不同细胞过程的子模型。图示用彩色单词表示这 28 个子模型，按类别分为代谢（橙色）、RNA（绿色）、蛋白质（蓝色）和 DNA（红色）。子模型通过常见的代谢物、RNA、蛋白质和染色体相互连接，分别用橙色、绿色、蓝色和红色箭头表示。B. 该模型通过 16 个细胞变量集成了细胞功能子模型。首先，模拟随机初始化到细胞周期的开始（左灰色箭头）；其次，对于每个 1 的时间步长（黑色箭头），子模型检索细胞变量的当前值，计算它们对细胞变量随时间变化的贡献，并更新细胞变量的值。在每次模拟的过程中，这种情况都会重复上千次。为了清晰起见，细胞功能和变量被分为五个生理类别：DNA（红色）、RNA（绿色）、蛋白质（蓝色）、代谢物（橙色）和其他（黑色）。变量和子模型之间的彩色线表示每个子模型预测的细胞变量。括号中指出了与每个子模型相关的基因数量。最后，模拟终止于细胞分裂，此时隔膜直径等于零（右灰色箭头）。来源：Cell, 2012, 150:389-401. https://doi.org/10.1016/j.cell.2012.05.044.

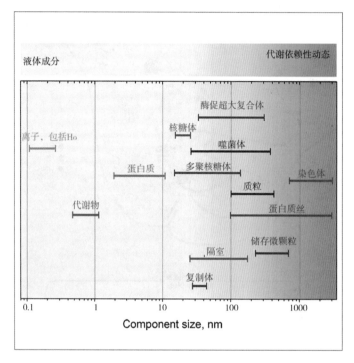

图 5.10　细菌细胞的分子大小和流动性取决于活跃的新陈代谢

来源：Cell, 2014, 156:183-194. https://doi.org/10.1016/j.cell.2013.11.028.

多物种的代谢模型可以通过体外实验和队列数据进行验证。多形拟杆菌本身可以产生乙酸和丙酸。对多形拟杆菌与青春双歧杆菌、溴瘤胃球菌，以及普氏栖粪杆菌或直肠真杆菌之一组成的 4 菌代谢模拟表明，由这 4 种菌组成的群落产生了丁酸和乙酸，而丙酸的含量明显低于多形拟杆菌单独存在时的水平[62]（见图 5.11，关于短链脂肪酸的更多资料，请参阅第 6 章，图 6.5）。文献 [63] 建立了微生物的丁酸生成模型，考察了 25 种产丁酸菌，确定了硫化氢对丁酸生成的抑制作用，以及 pH 对丁酸生成的影响等。

5.5　总结

人类共生菌群包括多个层级的分类单元，从生命的所有域到微生物菌株，以及单核苷酸多态性。通过参考基因组的积累和对每个身体部位的宏基因组装，分类工具可以为每个分类水平提供更加合适的标记基因序列。这样的进展将提高属和种级别的分类信息的准确性和多样性，这是确定传播路线（第 4 章）和疾病因果作用（第 6 章）的基础。可以从宏基因组组装的基因组中推断出潜在的功能，包括在特定培养基上的生长能力。低于物种水平的变异，通常被称为菌株，可以通过宏基因组测序或组织培养，也可以在同一个人身上长时间追踪。随着对共生微生物的体外特征化和多组学研究的不断深入，基于生物信息学和代谢模型的分类学将最终融合传统的功能分类学。

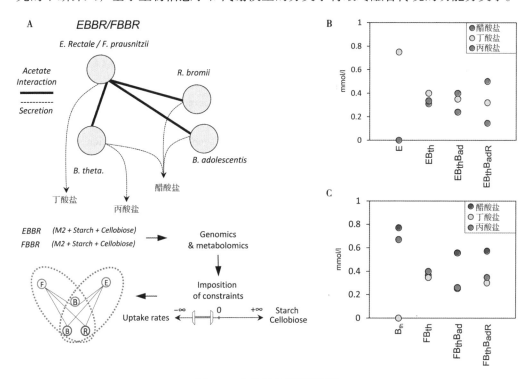

图 5.11　多物种代谢模型实例

A. 利用CASINO（群落和系统层面互作优化）工具箱，设计并模拟了硅微生物群落中的两个微生物群落：EBRR（隐蔽真杆菌+拟杆菌+缺陷双歧杆菌+反刍球菌）和FBBR（青春期粪便杆菌+青春期粪便杆菌+青春期粪便杆菌）。在添加了0.2%（质量/体积）淀粉和0.2%（质量/体积）纤维二糖的M2培养基中，比较了EBBR和FBBR群落的体外培养数据。CASINO中相互作用的细菌以及群落的表型通过一个优化算法进行识别。各菌株的生长均满足局部最优，而群落满足全局最优。利用群落的固定约束与计算出的动态约束的交点，通过局部力和群落力的求和得到群落的最优解。B.和C.网络结构对短链脂肪酸产生的影响。通过评估不同物种加入群落后短链脂肪酸的变化来检验CASINO优化的敏感性。首先，根据能力中心性（power centrality）和度数中心性对群落中最重要的受体（接受其他微生物代谢产物）和生产者（产生受体消耗的代谢产物）进行了鉴定。在模拟过程中，都添加了1mmol/L的葡萄糖，并预测了对应的SCFA曲线。在确定了主要受体和效应分子后，将其他种逐一添加到群落中，直至重建群落EBBR（B）和FBBR（C）。模拟之间的比较表明，SCFA在丰度和相互作用上对物种的缺少和存在十分敏感。来源：Cell Metab, 2015, 22:320-331. https://doi.org/10.1016/j.cmet.2015.07.001.

原著参考文献

[1] Bowers RM, Kyrpides NC, Stepanauskas R, Harmon-Smith M, Doud D, Reddy TBK, et al. Minimum information about a single amplified genome (MISAG) and a metagenome-assembled genome (MIMAG) of bacteria and archaea. Nat Biotechnol 2017;35:725–731. https://doi.

org/10.1038/nbt.3893.

[2] Vĕtrovský T, Baldrian P, Morais D. SEED 2: a user-friendly platform for amplicon high-throughput sequencing data analyses. Bioinformatics 2018;34:2292–2294. https://doi.org/10.1093/bioinformatics/bty071.

[3] Prodan A, Tremaroli V, Brolin H, Zwinderman AH, Nieuwdorp M, Levin E. Comparing bioinformatic pipelines for microbial 16S rRNA amplicon sequencing. PLoS One 2020;15. https://doi.org/10.1371/journal.pone.0227434, e0227434.

[4] Sun X, Hu YH, Wang J, Fang C, Li J, Han M, et al. Efficient and stable metabarcoding sequencing data using a DNBSEQ-G400 sequencer validated by comprehensive community analyses. Gigabyte 2021;2021:1–15. https://doi.org/10.46471/gigabyte.16.

[5] Qin J, Li Y, Cai Z, Li S, Zhu J, Zhang F, et al. A metagenome-wide association study of gut microbiota in type 2 diabetes. Nature 2012;490:55–60. https://doi.org/10.1038/nature11450.

[6] Sczyrba A, Hofmann P, Belmann P, Koslicki D, Janssen S, Dröge J, et al. Critical assessment of metagenome interpretation—a benchmark of metagenomics software. Nat Methods 2017;14:1063–1071. https://doi.org/10.1038/nmeth.4458.

[7] Meyer F, Fritz A, Deng ZL, Koslicki D, Gurevich A, Robertso G, et al. Critical Assessment of Metagenome Interpretation – the second round of challenges. bioRxiv 2021. https://doi.org/10.1101/2021.07.12.451567.

[8] Almeida A, Nayfach S, Boland M, Strozzi F, Beracochea M, Shi ZJ, et al. A unified catalog of 204,938 reference genomes from the human gut microbiome. Nat Biotechnol 2020. https://doi.org/10.1038/s41587-020-0603-3.

[9] Parks DH, Chuvochina M, Waite DW, Rinke C, Skarshewski A, Chaumeil PA, et al. A standardized bacterial taxonomy based on genome phylogeny substantially revises the tree of life. Nat Biotechnol 2018;36:996–1004. https://doi.org/10.1038/nbt.4229.

[10] Reimer LC, Vetcininova A, Carbasse JS, Söhngen C, Gleim D, Ebeling C, et al. BacDive in 2019: bacterial phenotypic data for high-throughput biodiversity analysis. Nucleic Acids Res 2019;47:D631–636. https://doi.org/10.1093/nar/gky879.

[11] Duran-Pinedo AE, Chen T, Teles R, Starr JR, Wang X, Krishnan K, et al. Community wide transcriptome of the oral microbiome in subjects with and without periodontitis. ISME J 2014;8:1659–1672. https://doi.org/10.1038/ismej.2014.23.

[12] Utter DR, He X, Cavanaugh CM, McLean JS, Bor B. The saccharibacterium TM7x elicits differential responses across its host range. ISME J 2020. https://doi.org/10.1038/s41396-020-00736-6.

[13] Zhu J. Over 50000 metagenomically assembled draft genomes for the human oral microbiome reveal new taxa and a male-specific bacterium; 2021. p. 2790.

[14] Coleman GA, Davín AA, Mahendrarajah TA, Szánthó LL, Spang A, Hugenholtz P, et al. A rooted phylogeny resolves early bacterial evolution. Science 2021;372. https://doi.org/10.1126/science.abe0511, eabe0511.

[15] Kempes CP, Wang L, Amend JP, Doyle J, Hoehler T. Evolutionary tradeoffs in cellular composition across diverse bacteria. ISME J 2016;10:2145–2157. https://doi.org/10.1038/ismej.2016.21.

[16] Mira A, Ochman H, Moran NA. Deletional bias and the evolution of bacterial genomes. Trends

Genet 2001;17:589–596. https://doi.org/10.1016/s0168-9525(01)02447-7.

[17] Probst AJ, Auerbach AK, Moissl-Eichinger C. Archaea on human skin. PLoS One 2013;8. https://doi.org/10.1371/journal.pone.0065388, e65388.

[18] Zou Y, Xue W, Luo G, Deng Z, Qin P, Guo R, et al. 1520 reference genomes from cultivated human gut bacteria enable functional microbiome analyses. Nat Biotechnol 2019;37:179–185. https://doi.org/10.1038/s41587-018-0008-8.

[19] Forster SC, Kumar N, Anonye BO, Almeida A, Viciani E, Stares MD, et al. A human gut bacterial genome and culture collection for precise and efficient metagenomic analysis. Nat Biotechnol 2019;37. https://doi.org/10.1038/s41587-018-0009-7.

[20] Groussin M, Poyet M, Sistiaga A, Kearney SM, Moniz K, Noel M, et al. Elevated rates of horizontal gene transfer in the industrialized human microbiome. Cell 2021;184:2053–2067.e18. https://doi.org/10.1016/j.cell.2021.02.052.

[21] Cross KL, Campbell JH, Balachandran M, Campbell AG, Cooper SJ, Griffen A, et al. Targeted isolation and cultivation of uncultivated bacteria by reverse genomics. Nat Biotechnol 2019. https://doi.org/10.1038/s41587-019-0260-6.

[22] Lagier JC, Dubourg G, Million M, Cadoret F, Bilen M, Fenollar F, et al. Culturing the human microbiota and culturomics. Nat Rev Microbiol 2018;16:540–550. https://doi.org/10.1038/s41579-018-0041-0.

[23] Truong DT, Franzosa EA, Tickle TL, Scholz M, Weingart G, Pasolli E, et al. MetaPhlAn2 for enhanced metagenomic taxonomic profiling. Nat Methods 2015;12:902–903. https://doi.org/10.1038/nmeth.3589.

[24] Milanese A, Mende DR, Paoli L, Salazar G, Ruscheweyh HJ, Cuenca M, et al. Microbial abundance, activity and population genomic profiling with mOTUs2. Nat Commun 2019;10:1014. https://doi.org/10.1038/s41467-019-08844-4.

[25] Ye SH, Siddle KJ, Park DJ, Sabeti PC. Benchmarking metagenomics tools for taxonomic classification. Cell 2019;178:779–794. https://doi.org/10.1016/j.cell.2019.07.010.

[26] Segata N. MetaPhlAn3; 2021. https://doi.org/10.1101/2020.11.19.388223.

[27] Weyrich LS, Duchene S, Soubrier J, Arriola L, Llamas B, Breen J, et al. Neanderthal behaviour, diet, and disease inferred from ancient DNA in dental calculus. Nature 2017;544:357–361. https://doi.org/10.1038/nature21674.

[28] Wibowo MC, Yang Z, Borry M, Hübner A, Huang KD, Tierney BT, et al. Reconstruction of ancient microbial genomes from the human gut. Nature 2021. https://doi.org/10.1038/s41586-021-03532-0.

[29] Rampelli S, Turroni S, Mallol C, Hernandez C, Galván B, Sistiaga A, et al. Components of a Neanderthal gut microbiome recovered from fecal sediments from El salt. Commun Biol 2021;4:169. https://doi.org/10.1038/s42003-021-01689-y.

[30] Fellows Yates JA, Velsko IM, Aron F, Posth C, Hofman CA, Austin RM, et al. The evolution and changing ecology of the African hominid oral microbiome. Proc Natl Acad Sci 2021;118. https://doi.org/10.1073/pnas.2021655118, e2021655118.

[31] Wood DE, Lu J, Langmead B. Improved metagenomic analysis with kraken 2. Genome Biol 2019;20:257. https://doi.org/10.1186/s13059-019-1891-0.

[32] Cao Y, Lin W, Li H. Large covariance estimation for compositional data via composition-adjusted thresholding. J Am Stat Assoc 2018;1–45. https://doi.org/10.1080/01621459.2018.1442340.

[33] Friedman J, Alm EJ. Inferring correlation networks from genomic survey data. PLoS Comput Biol 2012;8. https://doi.org/10.1371/journal.pcbi.1002687, e1002687.

[34] Scheiman J, Luber JM, Chavkin TA, MacDonald T, Tung A, Pham LD, et al. Meta omics analysis of elite athletes identifies a performance-enhancing microbe that functions via lactate metabolism. Nat Med 2019;25:1104–1109. https://doi.org/10.1038/s41591-019-0485-4.

[35] Deek RA, Li H. A zero-inflated latent Dirichlet allocation model for microbiome studies. Front Genet 2021;11. https://doi.org/10.3389/fgene.2020.602594.

[36] Liu X, Tong X, Zou Y, Lin X, Zhao H, Tian L, et al. Inter-determination of blood metabolite levels and gut microbiome supported by Mendelian randomization. BioRxiv 2020. https://doi.org/10.1101/2020.06.30.181438. 2020.06.30.

[37] Wang J, Jia H. Metagenome-wide association studies: fine-mining the microbiome. Nat Rev Microbiol 2016;14:508–522. https://doi.org/10.1038/nrmicro.2016.83.

[38] Zhang X, Zhang D, Jia H, Feng Q, Wang D, Di Liang D, et al. The oral and gut microbiomes are perturbed in rheumatoid arthritis and partly normalized after treatment. Nat Med 2015;21:895–905. https://doi.org/10.1038/nm.3914.

[39] Jie Z, Xia H, Zhong SL, Feng Q, Li S, Liang S, et al. The gut microbiome in atherosclerotic cardiovascular disease. Nat Commun 2017;8:845. https://doi.org/10.1038/s41467-017-00900-1.

[40] Jie Z, Liang S, Ding Q, Li F, Tang S, Wang D, et al. A transomic cohort as a reference point for promoting a healthy gut microbiome. Med Microecol 2021. https://doi.org/10.1016/j.medmic.2021.100039.

[41] Li J, Jia H, Cai X, Zhong H, Feng Q, Sunagawa S, et al. An integrated catalog of reference genes in the human gut microbiome. Nat Biotechnol 2014;32:834–841. https://doi.org/10.1038/nbt.2942.

[42] Browne HP, Forster SC, Anonye BO, Kumar N, Neville BA, Stares MD, et al. Culturing of 'unculturable' human microbiota reveals novel taxa and extensive sporulation. Nature 2016;533:543–546. https://doi.org/10.1038/nature17645.

[43] Darwin C. On the origin of species by means of natural selection, or the preservation of Favoured races in the struggle for life. London: John Murray; 1859.

[44] Mallet J. Darwin and species. In: Ruse M, editor. Cambridge Encycl. Darwin Evol. Thought. Cambridge: Cambridge University Press; 2020. p. 109–115. https://doi.org/10.1017/CBO9781139026895.013.

[45] Wallace AR. The method of organic evolution. Fortn Rev 1895;435–445. NS.57.

[46] Perlmutter JI, Bordenstein SR. Microorganisms in the reproductive tissues of arthropods. Nat Rev Microbiol 2020;18:97–111. https://doi.org/10.1038/s41579-019-0309-z.

[47] Brito IL, Yilmaz S, Huang K, Xu L, Jupiter SD, Jenkins AP, et al. Mobile genes in the human microbiome are structured from global to individual scales. Nature 2016;535:435–439. https://doi.org/10.1038/nature18927.

[48] Fraser C, Alm EJ, Polz MF, Spratt BG, Hanage WP. The bacterial species challenge : ecological diversity. Science 2009;323:741–746.

[49] Sheridan PO, Martin JC, Lawley TD, Browne HP, Harris HMB, Bernalier-Donadille A, et al. Polysaccharide utilization loci and nutritional specialization in a dominant group of butyrate-producing human colonic Firmicutes. Microb Genom 2016;2. https://doi.org/10.1099/mgen.0.000043, e000043.

[50] Rosenthal AZ, Qi Y, Hormoz S, Park J, Li SH-J, Elowitz MB. Metabolic interactions between dynamic bacterial subpopulations. Elife 2018;7. https://doi.org/10.7554/eLife.33099.

[51] Drake JW, Charlesworth B, Charlesworth D, Crow JF. Rates of spontaneous mutation. Genetics 1998;148:1667–1686.

[52] Zhao S, Lieberman TD, Poyet M, Kauffman KM, Gibbons SM, Groussin M, et al. Adaptive evolution within gut microbiomes of healthy people. Cell Host Microbe 2019. https://doi.org/10.1016/j.chom.2019.03.007.

[53] Jorth P, Staudinger BJ, Wu X, Hisert KB, Hayden H, Garudathri J, et al. Regional isolation drives bacterial diversification within cystic fibrosis lungs. Cell Host Microbe 2015;18:307–319. https://doi.org/10.1016/j.chom.2015.07.006.

[54] Kent AG, Vill AC, Shi Q, Satlin MJ, Brito IL. Widespread transfer of mobile antibiotic resistance genes within individual gut microbiomes revealed through bacterial hi-C. Nat Commun 2020;11:1–9. https://doi.org/10.1038/s41467-020-18164-7.

[55] Brito IL. Examining horizontal gene transfer in microbial communities. Nat Rev Microbiol 2021;1–12. https://doi.org/10.1038/s41579-021-00534-7.

[56] Baym M, Lieberman TD, Kelsic ED, Chait R, Gross R, Yelin I, et al. Spatiotemporal microbial evolution on antibiotic landscapes. Science 2016;353:1147–1151. https://doi.org/10.1126/science.aag0822.

[57] Karr JR, Sanghvi JC, Macklin DN, Gutschow MV, Jacobs JM, Bolival B, et al. A whole-cell computational model predicts phenotype from genotype. Cell 2012;150:389–401. https://doi.org/10.1016/j.cell.2012.05.044.

[58] Covert M. Simultaneous cross-evaluation of heterogeneous e coli datasets via mechanistic simulation. Science 2020. https://doi.org/10.1126/science.eaav3751.

[59] Bordbar A, Monk JM, King ZA, Palsson BO. Constraint-based models predict metabolic and associated cellular functions. Nat Rev Genet 2014;15:107–120. https://doi.org/10.1038/nrg3643.

[60] Mika JT, Poolman B. Macromolecule diffusion and confinement in prokaryotic cells. Curr Opin Biotechnol 2011;22(1):117–126. https://doi.org/10.1016/j.copbio.2010.09.009.

[61] Parry BR, Surovtsev IV, Cabeen MT, O'Hern CS, Dufresne ER, Jacobs-Wagner C. The bacterial cytoplasm has glass-like properties and is fluidized by metabolic activity. Cell 2014;156:183–194. https://doi.org/10.1016/j.cell.2013.11.028.

[62] Shoaie S, Ghaffari P, Kovatcheva-Datchary P, Mardinoglu A, Sen P, Pujos-Guillot E, et al. Quantifying diet-induced metabolic changes of the human gut microbiome. Cell Metab 2015;22:320–331. https://doi.org/10.1016/j.cmet.2015.07.001.

[63] Clark RL, Connors BM, Stevenson DM, Hromada SE, Hamilton JJ, Amador-Noguez D, et al. Design of synthetic human gut microbiome assembly and butyrate production. Nat Commun 2021;12:3254. https://doi.org/10.1038/s41467-021-22938-y.

第 6 章

共生微生物的疾病因果关系

摘　要： 对于那些在某些（尚未确定的）情况下可能引发疾病的正常菌群微生物，"条件致病菌"这个词并不是一个更合适的称呼。就像SARS-Cov-2引起的感染一样，要系统地理解致病性，需要考虑从人类和微生物遗传学到复杂疾病的病因学等众多因素。科赫法则需要根据当代因果推理理论中的证据标准进行修正。除了单一微生物的情况外，一个微生物可能需要其他微生物或人类细胞的分子才能表现出致病性。很多复杂疾病中，有致病性或者防病作用的微生物群落成员可以被确立。

关键词： 科赫法则，条件致病菌，因果关系，因果路径图，孟德尔随机化，糖尿病，口腔卫生

6.1　因果推断

1884年，罗伯特·科赫（Robert Koch）博士提出了现在被称为科赫法则的理论，用以判断一种细菌是否为某种疾病的病原体（图6.1、图6.2）[1,2]。把致死性疾病归因到一种细菌，这在当年是很不容易的。1859年，巴斯德用天鹅颈烧瓶实验证明了肉汤的变质是由空气中的细菌（$10^4 \sim 10^6$ CFU/m^2[3]，与第3章所示的肺微生物组相比较，见图3.6）引起的，而不是肉汤中自发生成的生命体。此外，列文虎克早期对口腔细菌的研究（第1章，示例1.1）也显示了细菌是无处不在的，而且大多数是无害的，是日常生活的一部分。

在科赫的时代，由于还没有宏基因组学技术，因此不论是在宿主动物还是在培养物中，显微镜成为区分不同种细菌的金标准（图6.1）。因此，在那个人们不常洗手的年代，如果要把一种细菌和炭疽这样的致命疾病联系起来，他采用了最严格的证明。巴斯德从预防角度归因，还开发了一种预防炭疽热的疫苗。罗伯特·科赫博士的创新性实验则涉及用分离出来的活细菌（或孢子）引起疾病，并进一步在显微镜下证明引起疾病的仍然是这种细菌。细菌和疾病的关联不必非黑即白，单菌干预的致病性也不必100%成功（图6.1、图6.2），使那些已经夺去了许多生命的著名病原体并没有机会满足科赫法则。

第 6 章　共生微生物的疾病因果关系

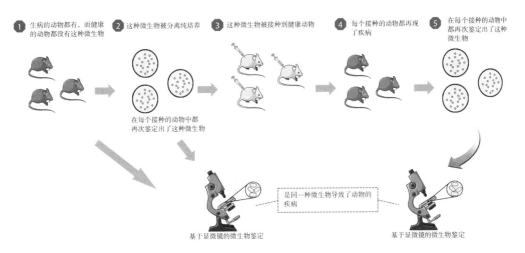

图 6.1　科赫对炭疽杆菌的实验

这些步骤是根据科赫法则的要求（图6.2）以及当时的技术——显微镜和分离培养。健康的动物没有显示，因为根据科赫法则，其中不应该有微生物。

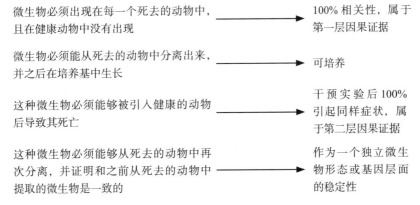

图 6.2　科赫法则及其与现行因果推断体系的对应

科赫[1,2]被分解为关联性——第一级的因果证据和干预性——第二级的因果证据。对于分离培养和稳定的重新隔离的要求，更多的是出于分类学和进化学的考虑。只要我们对测序鉴定的微生物有一个恰当的命名（并记录使用的参考基因组），分离培养就与因果推理无关，但是对于机制和治疗研究仍然是有益的。重新隔离排除了另一种可能的假设，即一些微生物无意中被当作病原微生物的无害化动物（图中未示）感染了同样的疾病。因此，是否患病需要被准确鉴定，实验室动物应在标准护理下，以确保重复性。

示例 6.1　奥卡姆剃刀原则或牛顿的简单性原理

埃德温·汤普森·杰尼斯教授在他去世后出版的《概率论沉思录》[5]一书的第 20 章中阐述了奥卡姆剃刀的概率论本质。"不要引入无助于提高推论质量的细节。"奥卡姆剃刀原则由圣方济会修道士威廉·奥卡姆于1330年提出，它声称"如无必要，勿增实体"，也就是说，"能用更少的假设完成的事，引入更多就是徒劳"。

125

奥卡姆剃刀主要通过考虑先验信息[5]，对某个模型进行惩罚。正统的统计理论只是用"抽样分布"来比较模型。根据贝叶斯定理，如果数据与先前的信息相比信息量很大，那么两个模型的相对优劣就取决于在各自的参数空间上达到的似然性有多大，以及有多少先验概率集中在各自的高似然区域。对于一个信息量相对较大的实验，我们预计其概率空间可能集中在相对小的一个区域，而一个"更简单"的模型将占据更小的参数空间，因此奥卡姆剃刀原则更偏好这个更合理的模型[5]。

此外，杰尼斯教授在他的书的第7章中证明，如果我们没有任何信息，正态分布是最合理的估计。但是，当我们确实有先验信息时，我们不应该假装我们没有。

在1687年出版的《自然哲学的数学原理》一书中，牛顿总结了四条推理规则。其中第一条规则——"对自然现象的解释不应增加非必要的因素，除非它们是真实的且能充分解释自然现象"也强调了简单性的原则，即不允许引入过多的复杂性（参数），因为这些复杂性并不能提高解释的真实性。这也是爱因斯坦在创立相对论时所遵循的方法，他将牛顿时代所不知道的证据纳入了考虑。

生物医学领域（以及金融领域）一直深受费希尔爵士提出的统计原理的影响[5,6]。然而，物理学家们知道，真正重要的不是样本大小或 P 值，而是各种可能的场景，以及每个场景的概率与现有证据的关系。这些概率会随着新的证据而更新。例如，巴斯德最终证明了肉汤变质"自然发生"理论是错误的，降低了其他解释的可能性[4]。如果出现了一种新的解释，那么可能性就需要根据不同的情景中重新分配，而且一些证据可能更支持新的解释而非旧的解释。当然，有些人会比其他人更难接受新的解释[5]。回到科赫法则，每个实验都排除了哪些其他的解释？我们是否越来越相信炭疽杆菌是炭疽热的原因（图 6.1，图 6.2）？

找到替代的解释（假说）和判别每种解释可能性的能力可以依赖于我们自己的先验知识。例如，人类微生物组计划中一项对美国阴道菌群的分析发现，志愿者是否有大学学历与乳酸杆菌和非乳酸杆菌阴道菌群类型有关。这里可能的解释有哪些？（第8章将详细介绍宫颈及阴道微生物组的相关内容）。

思考题 6.1

美国肠道项目（16S rRNA 扩增子测序）最近发表了一篇文章，发现在控制了酒精摄入量和饮食信息后，与 2 型糖尿病有关的大部分粪便微生物组生物标志物都不显著了[12]。

膳食信息可归结为营养成分，如氨基酸、六胺、纤维[13]、糖、脂类和添加剂（在流行病学上，酒精被认为是类风湿性关节炎的保护因素[14]）。急性乙醇摄入会导致外周血淋巴细胞 MHC Ⅰ 类表达增加[16]。乙醇及其代谢产物乙酸（图 6.5，第 6.3.1 节），强有力地调节 T 滤泡辅助细胞（Tfh）的功能，并促进醋酸盐诱导的 IgA 的产生。

在不迷失于细节的情况下，如何用因果路径图来描述因果假设？从你更熟悉的糖尿人群入手，你认为哪条因果路径对患有 2 型糖尿病人群更重要？你还会寻找其他的证据吗？

"混杂因素"是一个来自统计学的模糊术语，应该被抛弃[5,6]。虽然在一项研究的开始可能存在许多未知的因素，但是为了使分析在数学上成立，每个场景仍然需要路径图来清晰地表示。试图控制（保持不变的）每一个因素，如果它们是对撞因子，将导致错误的负相关；如果它们是因果路径的一部分，将导致无法检出相关性[6]。例如，如果控制了粪便中栖粪杆菌属的丰度，即通过它的因果路径被废用，因此只能是在寻找其他可能有贡献的菌，再去研究碳水化合物（如面条）的摄入对快乐情绪的影响（图 6.3）；或者在研究肺癌时控制吸烟，也是废用了关键因果链条。关于对撞因子，一个常见的例子是"帅气但是渣的家伙"——显而易见的负相关，因为又渣又不帅的家伙更加没有机会，或者才能和美貌在演员中不可兼得。一个更相关的例子（图 6.3）是：如果我们都认为吃得太多或锻炼得太少会导致肥胖（更高的腰臀比），那么在分析中控制腰臀比或 BMI 会让吃得太多和锻炼得太少之间出现负相关，也就是说，由

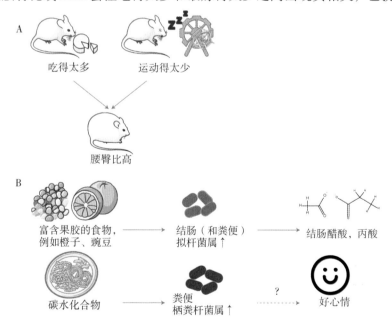

图 6.3 几个因果路径图的例子

说明了为什么在分析中对某些变量进行统计控制（固定其值）会导致错误的结论。A. 碰撞因子。B. 在同一条因果路径上。粪大肠埃希菌和精神健康之间的关系目前属于相关性。

于控制了对撞因子（固定的腰臀比或体重），饮食和运动之间会呈现正相关。因此，我们会陷入饮食、锻炼和更多饮食的循环中，同时试图保持相同的体型，而如果不通过保持体型一致这扇门，吃得多和运动得少本身并没有负相关。

通过描述当前的假设，因果路径图[5,6]为进一步的研究提供了一个透明的公共基础。例如，当我们讨论剖宫产和阴道分娩的婴儿的微生物组的差异，或者早产和足月分娩的差异时，我们是否测量了必要的激素？母乳中含有的皮质类固醇不仅在呼吸系统发育中发挥作用[9]，还已经在大鼠和小鼠中被证明能够启动下丘脑－垂体－肾上腺皮质（HPA）轴[10]，而这是应激反应的核心[11]。我们总是可以添加新的路径以获得更加完整或更加精确的描述。当水手们只知道柠檬可以预防坏血病时，他们就可能把柠檬煮沸（这会破坏维生素C），人们因此就不再相信柠檬的功效了[6]（图6.4）。

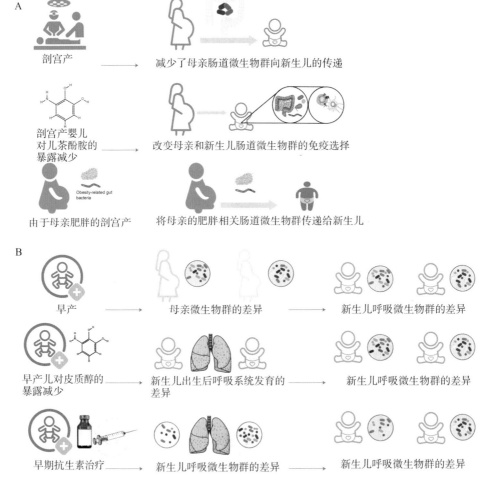

图6.4 可以解释婴儿微生物组差异的替代路径图的例子

A. 剖宫产的影响。B. 早产的影响

6.2 人类微生物和疾病之间关系的当前证据等级

目前已有大量的研究探讨了正常菌群中的不同分类群与疾病之间的关系，以及微生物组中的不同分类群与疾病相关的循环分子，如细胞因子、脂类、氨基酸的关联。然而，门水平的结果往往受到相对丰度这一成分数据性质的影响[20]。例如，厚壁菌门和拟杆菌门在小鼠肠道中的总数几乎等于1，所以它们看起来呈负相关，任何厚壁菌门的丰度较高的疾病会显示为拟杆菌门丰度较低。肠道等部位宏基因组属或种水平的数据已经包含数百个分类单元，且相对丰度分布更均匀，成分数据这总和为1的约束基本不需要另外变换处理（不像阴道样本中乳酸杆菌含量大于90%）[20,21]。

除了相关性，某些微生物和疾病的关系达到了证据等级的第二层——（随机）干预（示例6.2）通过改变暴露水平来展示其影响，或第三层，反事实，通过在头脑中模拟不进行暴露来检验因果性（表6.1），这是目前AI无法做到的。

示例 6.2 孟德尔随机化（MR）

在干预研究中，我们常常面临这样一个问题：当我们试图观察变量 X 对结果 Y 的影响时，其他因素可能会影响 X。例如，每天喝酸奶的习惯可能与一个人的工作、教育、经济等方面状况有关，这也可能与定期去健身房有关。这并不是说我们应该放弃纵向队列研究，如果几十年来煞费苦心地进行，它们会有独特价值，将有助于反事实推断。

随机对照试验（RCTs）则是通过随机的方式分配 X 的值，从而切断了它与其他因素的联系，例如，我们可以随机地让一部分人喝酸奶，另一部分人不喝酸奶，而这与他们的其他情况没有任何关联，这样，我们可以更有信心地观察酸奶是否对心血管健康、胃肠健康有益。

孟德尔随机化也是在考察 X 对 Y 的因果效应时隔绝其他因素对 X 的影响的有效方法。孟德尔随机化利用了减数分裂产生配子时染色体发生的随机分配。虽然单一 SNP 的随机更易理解，目前往往需要采用多个与 X 相关的 SNPs 作为工具变量。这些 SNPs 必须能够解释 X 的较大比例的方差（例如，大于20%），否则我们就很难检测到 X 对 Y 的影响。此外，这些 SNPs 还必须满足两个条件：一是它们不能直接影响结果变量 Y，而只能通过 X 间接影响；二是它们不能与另一个同时作用于 X 和 Y 的因素关联——那这个因素也是因果路径的一部分，而不是所谓"混淆因素"。

孟德尔随机化在发现因果关系方面非常强大，尤其是在那些随机对照实验由于伦理原因不可行的情况下。例如，通过两样本的 MR 结合来自中国 4D-SZ 队列的 SNPs-微生物组关联分析和来自 Biobank Japan 的 SNPs-疾病关联分析（BBJ），肠道中的副血链球菌 *Streptococcus parasanguinis* 已被发现和心脏后壁厚度这一问题，以及大肠癌有因果关联，提供了超过宏基因组关联分析研究能得出的因果证据水平[23-25]（表 6.1）。另外，MR 检测到高 BMI、吸烟和饮用咖啡可以增加类风湿性关节炎风险，而铁、亚油酸（一种主要的促进睾丸激素合成的多不饱和脂肪）和受教育程度具有保护作用[26]。

当解读来自大队列的研究结果时，必须注意，我们的分析受到表型和调查问卷收集的限制，因此，即使发现了一个因果信号，仍然可能意味着另一个未被问及的问题，该问题至少应当与现有的问题在同一条因果路径上。

我们鼓励读者阅读朱迪亚·珀尔教授的为普通人撰写的《为什么》一书，其中有许多来自其他学科的例子[5,6]。如果没有随机对照试验，孟德尔随机化可以使用人类遗传信息（摘要统计数据，不需要个人遗传数据），或使用来自同一个队列的所有测量值（单样本和两样本 MR）来提供第二级的因果证据（示例 6.1）[22,26]。对小鼠的干预实验通常不是随机的，但与人类不同，我们没有特定的理由相信这某只小鼠被分配到实验组是受到小鼠的某些表型的影响（比如更胖的小鼠对干预更配合？）。但是无菌小鼠的代谢、免疫和神经系统状态都不正常[40-42]。无特异性病原体（SPF）小鼠和人类之间，以及小鼠之间的微生物组在种类和互作方面也有很大的不同，这对于物种水平的关联和干预都适用[43,44]。

直觉地，与仅对健康对照组与疾病组的微生物进行关联分析相比，涉及导致疾病的已确定的分子的微生物关联被认为具有更强的可信度（表 6.1）。

表 6.1 微生物研究中因果关系的例子

因果证据等级	证据	参考文献
1 相关性	粪便大肠埃希菌与 GLP-1 水平相关；与对照组相比，糖尿病前期患者的粪便大肠埃希菌更为丰富	[27,28]
1 相关性	与对照组相比，动脉粥样硬化性心血管疾病患者粪便中的大肠埃希菌含量较高，并且与手握力呈负相关，手握力是已知的心血管事件的流行病学因素	[24,29]
1 相关性	在 2 型糖尿病患者中，粪便拟杆菌（*Bacteroides caccae*）富集	[28,30]
1 相关性	粪便爱格氏菌在糖尿病早期患者中富集，且和虚弱相关	[28,31,32]
1 相关性	粪便中扭瘤胃球菌在溃疡性结肠炎患者中增多，并且根据 BSS 得分与粪便松散程度相关	[33,34]
1 相关性	吸烟与口腔中韦荣氏菌和普雷沃菌的关联	4.41 节
1 相关性	口腔 *Lachnoanaerobaculum umeaense*、*Oribacterium* 和与血清尿酸水平及尿酸转运蛋白基因 SLC2A9 上的 SNP 相关	[35]

续表

因果证据等级	证据	参考文献
2 干预（RCT）	对乳酸杆菌和双歧杆菌［9种混合，包括长双歧杆菌、短双歧杆菌、干酪乳杆菌、卷曲乳杆菌、发酵乳杆菌、植物乳杆菌、鼠李糖乳杆菌、唾液乳杆菌和格氏（加氏）乳杆菌］进行多中心随机对照实验，配合小檗碱治疗（抗生素庆大霉素硫酸盐治疗后），T2D 患者的血糖水平并没有改善，但对血脂有一定影响	[36]
2 干预（RCT）	治疗后 12 周内卷曲乳杆菌（CTV-05）的随机对照试验减少了细菌性阴道炎的复发（46/152 复发，对照组 34/76 复发）	[37]
2 干预（MR）	粪便中的毛螺菌与血清尿酸水平互为因果，粪便中的降解果胶的微生物（例如来自拟杆菌属或梭杆菌）可增加血清尿酸水平	[25]
2 干预（MR）	粪便大肠埃希菌可能导致 T2D、心脏衰竭、结肠癌等	[25]
2 干预（MR）	粪便中副血链球菌可能导致心脏后壁变厚和大肠癌	[25]
2 干预（MR）	粪便中的 Saccharibacteria（TM7）可以降低血清肌酐，增加以肾小球滤过率预估的肾功能	[38]
3 反事实	如果没有人类乳突病毒，就不会有宫颈癌	针对每次检测，可提供免费的保险
3 反事实	如果没有幽门螺杆菌，是否就不会有胃癌？是否就不会有胃溃疡？	
3 反事实	如果没有牙龈卟啉单胞菌、齿状密螺旋体、福赛斯坦纳菌和中间普雷沃菌，是否就没有牙周炎？	

这并不是一个详尽的列表，而且相关性的研究特别丰富。该表旨在展示《为什么》一书中提到的因果证据的三个层次在微生物领域中的应用。

如果在同一条因果路径上发现了更多的相关性，那么这关联是假阳性的概率就会大大降低（Jaynes 教授的书[5]在数学上可能令许多读者望而却步，但是你可以放心地跳过一些方程）。宏基因组 – 全基因组关联研究（M-GWAS）与人类基因的关联，在证据等级上可能比仅有微生物与疾病关联显得略高（表 6.1），因为人类基因与疾病之间已经建立的关联，增加了微生物与疾病之间新的关联的可信度，缩小了参数空间，并指明了需要做的实验。关联分析的优势在于，我们并不是盲目地遵循一个假设，而这个假设可能最终只是大局中的一个小问题，有关联总比没有关联强。

幽门螺杆菌是一种胃病原体，虽然在发达国家通常建议根除，但仍然感染了全球一半的人口[45]。根据幽门螺杆菌的基因特征、寄生位置、自身的基因和免疫反应以及可能存在的保护型乳酸菌等因素，只有 1% ~ 3% 的幽门螺杆菌感染者[46]会发展成胃癌[45-47]。幽门螺杆菌作为十二指肠和胃溃疡以及胃癌的最大危险因素，还没有达到因果证据的三级，不是因为不到 100% 的疾病表现（科赫法则，图 6.2）。不完全发病并不意味着我们不能确定幽门螺杆菌与胃癌之间的因果关系，而是意味着还有其他因素参与了因果路径图。我们仍然可以根据反事实推断（表 6.1）告诉人们，如果没有幽门螺杆菌的感染，他们在未来几年内不会得胃癌。如果事实证明我们必须排除一些不是由幽门螺杆菌引起的胃癌（例如具核梭菌可能也引起胃癌），这仍然是一个反事实推断。胃癌患者并不常规检查幽门螺杆菌的感染情况，对于一些患者来说，可能是由于幽门螺杆菌的根除或自发消除（例如，饮用咖啡有助于增加乳酸杆菌？这一说法还没有得到证实）来得太晚了，以至于无法逆转对胃黏膜的损害，这一点还需要进一步研究。

促牙周炎的细菌通常会通过机械或化学的方式去除或消灭（例如，针对牙龈卟啉单胞菌的漱口水）。如果我们对这些细菌在牙周炎中的因果关系进行反事实推断，并给出肯定的回答（表 6.1），那么我们应该采取什么样的共识性做法呢？这些细菌是否对宿主的免疫力有正向的影响（例如，抵御不再具有威胁性的病原体）或对宿主的新陈代谢有有益的作用（例如，唾液卟啉菌与人类基因组中淀粉酶基因的高拷贝数相关[48]）？从微生物学的角度来看，用于口腔卫生的化学物质是否有副作用？除了牙周炎和龋齿外，如果口腔微生物组中含有过多的结肠直肠癌、结肠炎或促肝病的细菌（例如，参见 [49-51]），即使这些微生物还没有出现在粪便中，我们又打算如何监测或干预呢？

微生物群落是复杂的生态系统，其中一些微生物会改变或提供公共服务（非特异性互惠）给它们的栖息地，从而影响其他许多微生物（第 2 章）。例如，乳杆菌和双歧杆菌能将亚硝酸盐还原为一氧化氮，这不仅可以调节血液流量和血液流动性，还可以消除潜在的致癌物亚硝酸盐[52]（这对于幽门螺杆菌和胃癌的全面分析很有意义）。活泼瘤胃球菌是一种常见的克罗恩病相关的细菌，它能将 ABO 血型的 B 型糖链水解为 O 型糖链[53]，其降解黏蛋白聚糖的方式与其他生物蛋白不同[54]，并能分泌一种复杂的葡聚糖聚糖，通过 TLR4 受体诱导产生炎症反应[55]。在定菌小鼠中，已经证明来自拟杆菌属的唾液酸酶可以从肠道黏膜中释放出唾液酸，而这种唾液酸是一些病原菌，如沙门菌、难辨梭状芽孢杆菌（原名艰难梭菌）增殖所需的[56]。阴道中的阴道加德菌可以表达唾液酸酶；二路普雷沃菌编码硫酸酯酶和唾液酸酶，这两种酶会破坏健康的黏液层，导致菌群失调，进而引起细菌性阴道炎，甚至远期促成早产[57,58]。肠道白色念珠菌是一个主要的诱导 Th17 细胞反应的真菌，诱导后的细胞反应将通过交叉反应会对抗其他真菌，如会导致呼吸道炎症的烟曲霉[59]。在动物模型中，表皮葡萄球菌、藤黄微球菌或藤黄微球菌细胞壁肽聚糖共同存在降低了金黄色葡萄球菌致病剂量，这是由于共栖菌存在减少了肝脏巨噬细胞的氧化爆发[60]。目前，一些认为单一微生物导致疾病的结论可能只是更复杂的因果路径图的一部分。我们最终将理解每条路径机制和概率，并为每个患者做出明智的决定（第 7 章）。

6.3　从微生物到分子

类比新冠病毒感染和症状的对应关系，本节系统地解释微生物的致病性需要涵盖从人类和微生物遗传学到复杂疾病的病因学的所有内容。一种微生物中的多种代谢分子都可能在疾病中起到致病或有益的作用。

6.3.1 嗜黏蛋白阿克曼菌的多种有效分子

嗜黏蛋白阿克曼菌（*Akkermansia muciniphila*）是一种与小鼠体重[61]、粪便盐度[62]呈负相关的微生物，在Roux-en-Y胃旁路术（RYGB）小鼠模型[63]和二甲双胍治疗的T2D病人中[64]丰度升高。它的外膜蛋白Amuc_1100能够通过TLR2与宿主细胞通信，这一蛋白部分解释了加热杀死的嗜黏蛋白阿克曼菌在小鼠实验中效果不比活菌差。一项对32例接受巴氏消毒后嗜黏蛋白阿克曼菌志愿者进行了单中心随机对照实验显示，患者的胰岛素敏感性明显提高，体重、脂肪量和臀围也有一定改善[66]。

另外，相比于某些厚壁菌，嗜黏蛋白阿克曼菌在低温环境下丰度相比拟杆菌门降低[67]，而生热作用会增强[68]。这一功能与其分泌的一种名为P9的蛋白质有关，这种蛋白质可以诱导胰高血糖素样肽-1（GLP-1）的分泌，从而在循环中促进葡萄糖的内稳态，并诱导解偶联蛋白1（UCP1）在褐色脂肪组织中产热[68]。嗜黏蛋白阿克曼菌还能够与其他许多细菌一起产生乙酸盐（一种主要的短链脂肪酸，图6.5），能够增进食欲（类似醋的作用），促进滤泡辅助T细胞，并诱导产生更多的IgA。

一项MR研究报道了粪便中的丙酸会促进2型糖尿病的发生[70]，这可能是由于拟杆菌属和普雷沃菌属的多效作用，这两类细菌既能产生丙酸盐（参阅第6.3.2节和6.3.3节），又能产生较少的琥珀酸促进小肠的糖异生[71]，从而促进糖尿病的发生。在鸡的十二指肠中，嗜黏蛋白阿克曼菌的丰度与饲料效率相关，盲肠中乳杆菌属、类杆菌、棒状杆菌的丰度也与饲料喂养的效率相关[72]。

正如在第2章中所提到的，阿克曼菌也与一些疾病有关，如大肠癌[73,74]、动脉粥状硬化心血管疾病[24]、阿尔茨海默病和精神分裂症[75,76]。疾病的关联可能涉及细菌的一些不同的基因组特征[77,78]，是否能够接触一些不同的宿主分子（如感染和癌症病例中表达的muc2以外的黏蛋白[79]）、昼夜节律（第2章，示例2.3）或其他一些机制。

回到膜蛋白，来自Amuc_rs03735和阿克曼菌的Amuc_rs03740的T细胞相互作用肽特异性诱导了IgG$_1$，而不是更常见的肠道IgA[80]。这种免疫调节作用又增添了对于把阿克曼菌用于PD-1肿瘤免疫治疗的辅助治疗的信心。

6.3.2 支链氨基酸对增肌和糖尿病的促进作用

所有*Prevotella copri*复合体分支都产生支链氨基酸（BCAAs）[81]。在这种情况下，尽管除了普雷沃菌的其他细菌也参与了支链氨基酸的产生和代谢，但普雷沃菌对支链氨基酸的产生有因果关系，并且对支链氨基酸已经确定有影响的疾病可能有贡献。

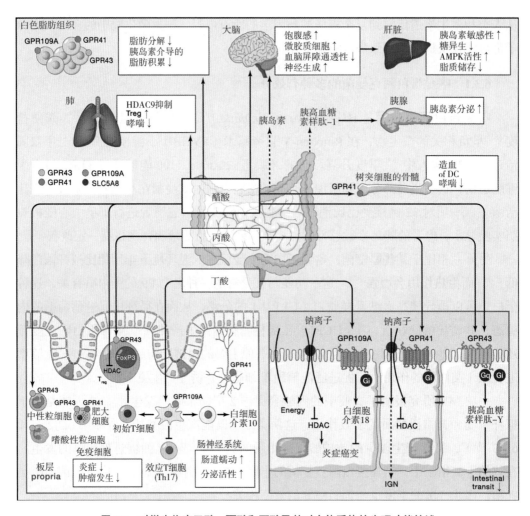

图 6.5 对微生物产乙酸、丙酸和丁酸及其对人体受体的生理功能综述

微生物通过多种生化途径将膳食纤维的发酵成SCFAs。字母的大小代表不同SCFAs的相对比例。在远端肠道，SCFAs可以通过扩散或SLC5A8介导的转运通道进入细胞，作为能源或组蛋白去乙酰化酶（HDAC）抑制剂。GPR41和GPR43受体感知乙酸或丙酸腔内浓度，释放PYY和GLP-1，影响饱腹感和肠道转运。丁酸内酯通过抑制GPR109A和HDAC发挥抗炎作用。此外，丙酸可以通过IGN转化为葡萄糖，增加饱腹感和减少肝脏葡萄糖的产生。SCFAs还可以作用于肠道内的其他部位，如肠神经系统（ENS），刺激肠道运动和分泌活动，或固有层中的免疫细胞，减少炎症和肿瘤发生。少量的SCFAs（主要是乙酸盐和丙酸盐）进入血液循环，能直接影响脂肪组织、大脑和肝脏，诱导对机体整体有益的代谢效应。实心箭头表示每种SCFAs的直接作用，虚箭头表示来自肠道的间接作用。关于疾病，请参阅文献[17]。来源：Cell, 2016, 165:1332-1345. https://doi.org/10.1016/j.cell.2016.05.041.

在中老年人群中，T2D是支链氨基酸代谢紊乱的主要下游疾病[39,82,83]。在年轻时，支链氨基酸有助于肌肉生长[84,85]。植物性和动物性食物中的支链氨基酸含量差异促成了低蛋白或素食饮食对预防T2D的有益效果[86]。除了肌肉、心脏、组织和肝脏的功能外，支链氨基酸还是雷帕霉素（mTOR）通路的激活剂。维持处于高代谢状态的调节性T细胞需要该通路[87]，这可能解释了社会经济状况与自身免疫性疾病（如类风湿关节炎等）之间的联系。

正如常用的对 SCFAs（短链脂肪酸）这一缩写的使用，BCAAs 可能是另一个模糊了分子之间差异，造成混淆的缩写。减少异亮氨酸或缬氨酸，而不是亮氨酸的摄入，最近已被证明可以改善代谢健康。饮食水平的异亮氨酸与 BMI 有着显著的相关性[86]。

6.3.3 自身抗原的分子拟态

除普雷沃杆菌外，另一种主要的"肠型"（第2章），拟杆菌属，与1型和2型糖尿病、大肠癌、炎症性肠病有关[28,30,31,39,74,88-90]。除了 SCFAs、BCAAs 和 LPS（脂多糖）结构外[91]，拟杆菌蛋白质的抗原特性也被发现在疾病中发挥作用。多氏拟杆菌的整合酶（一种通过转座子的携带的酶，可以将自身整合到基因组中），包含一个低亲和力的胰岛β细胞特异性葡萄糖-6-磷酸酶-催化亚单位相关蛋白（IGRP，氨基酸206-214）的片段，它能够诱导糖尿病细胞毒性 T 细胞出现在肠道，并以 MHC-Ⅰ特异的方式抑制结肠炎的发生（图6.6）。某些人的胰岛β细胞可能表达了过多的易感 MHC-Ⅰ，而胰岛β细胞代替了自身抗原负载的树突状细胞（DC）在胰淋巴结（PLN）中被杀死（这样分泌胰岛素的细胞就少了）。

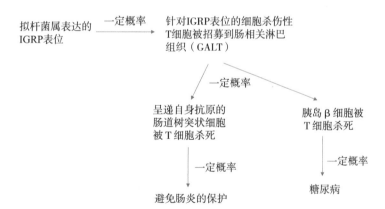

图 6.6　拟杆菌整合酶自身抗原路径图的可能模型

来自文献[92]中的总结。每一步的可能性取决于细胞种群及其进化历史。有些人可能永远不会得这种疾病。

T 细胞受体（TCRs）能够识别多形拟杆菌（*Bacteroides thetaiotaomicron*，BT4295，氨基酸541-554）的外膜多糖化合物（PUL）蛋白中的另一个肽段，该肽段没有已知的拟态，也能够通过小鼠模型中调节性 T 细胞对结肠炎起到保护作用[93]。BT4295 包含在外膜囊泡（OMVs），而 BT4295 的表达受到膳食葡萄糖的抑制[93]。由于对炎症和代谢性疾病的易感性可能不同，因此需要将拟杆菌肠型进一步细分为诸如多形拟杆菌和单形拟杆菌等亚型。

6.3.4 其他例子，外膜小泡，噬菌体

大肠埃希菌是一种在糖尿病前期个体中富集的微生物，它能够产生吲哚（表6.1），这是一种能够刺激肠内分泌 L 细胞释放 GLP-1 的物质，从而使胰岛 β 细胞分泌胰岛素[98]。因此，在 RYGB 小鼠模型中[63]，大肠埃希菌以及嗜黏蛋白阿克曼菌的丰度增加，很可能是手术治疗代谢综合征疗效的关键之一。

当然，大肠埃希菌还具有许多其他功能。在胰腺癌中，由大肠埃希菌编码的胞苷脱氨酶（CDD，长亚型）可代谢庆大霉素，2′, 2′- 二氟脱氧胞苷（图 6.7）[99]。通过生物信息学分析和体外实验，我们可以更好地了解微生物组中的个体成员的作用和机制。

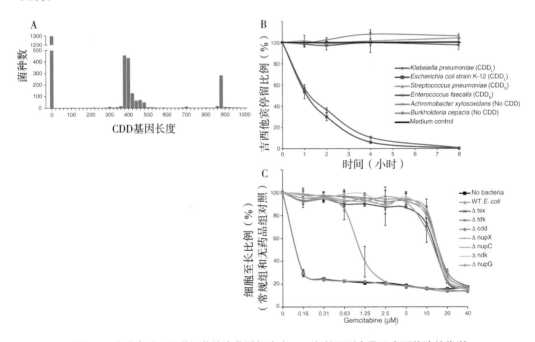

图 6.7　实验表明，肠道细菌的胞苷脱氨酶（CDD）长亚型介导了吉西他滨的代谢

A. KEGG数据库中所有细菌的CDD DNA序列及其对应细胞数量长度直方图。B. 吉西他滨（4mm）与107株细菌在M9微盐培养基中共培养。通过过滤法分离细菌，用HPLC-MS/MS检测培养基中剩余的吉西他滨的含量。纵轴表示两个复制品之间的标准差，每个复制品包含两个重复实验。C. 野生型亲本大肠埃希菌K-12株（长CDD）、CDD敲除株（△CDD）和无菌培养基分别与不同浓度的吉西他滨孵化4h，然后筛选出细菌，将流通培养基添加到GFP标记的ASPC1人胰腺癌细胞中。7d后，用绿色荧光蛋白（GFP）检测ASPC1细胞的生长情况，与无药物控制组进行比较。条形代表4个复制之间的标准差。来源：Science, 2017, 357:1156-1160. https://doi.org/10.1126/science.aah5043.

在肺组织中，卵形拟杆菌、粪便拟杆菌和黑色素原普雷沃菌能够释放外膜囊泡，这些囊泡能够激活肺泡巨噬细胞，使其产生 IL-17B，从而促进肺纤维化[100]。

OMVs 的产生并不是革兰氏阴性细菌的专利，革兰氏阳性菌也可以释放膜囊

泡[101]。例如，阴道卷曲乳杆菌（*L. crispatus*）BC3 和格氏乳杆菌（*L. gasseri*）BC12 能够产生直径大于 100nm 的囊泡，降低了 HIV-1（human immunodeficiency virus-1）对靶细胞的黏附力，防止病毒侵入[102]。这种 OMVs 介导的病毒防御机制，是阴道乳酸杆菌除了通过产生乳酸和过氧化氢来抵抗病毒的另一种途径，也是它们与宿主细胞相互作用的方式之一。

对于酒精性肝病，粪便微生物组中有溶细胞素的患者死亡率更高（某些粪肠球菌菌株表达的溶细胞素）[103]。小鼠体内实验发现针对粪肠球菌的噬菌体可以减少酒精引起的肝损伤和脂肪变性[103]。

6.4 总结

宏基因组学是一个数据驱动的研究领域，它发现了许多与微生物组相关的现象，这些相关性是通向因果关系的 I 级证据。随机对照试验和大人群孟德尔随机化分析为某些相关性提供了 II 级证据（表 6.1）。前瞻性队列研究和更多关于微生物组的基础研究将增强我们对第三级证据的信心，这些证据将指导公共卫生决策。如果我们能够像对一些传统的病原体那样，深入了解正常菌群的成员及其产物，那么即使分子机制和相互作用再复杂，因果关系也会清晰明了。然而，我们必须按照一定的优先级进行研究。

原著参考文献

［1］Tortora GJ, Funke BR, CL CT, editors. Chapter 14 Principles of disease and epidemiology copyright. In: Microbiol. An introd. 10th ed. Pearson; 2010.

［2］Tu A-HT. 15.2 How pathogens cause disease—microbiology | OpenStax. In: Parker N, Schneegurt M, Lister PM, Forster B, editors. Microbiology. The American Society for Microbiology Press; 2016.

［3］Dickson RP, Erb-Downward JR, Martinez FJ, Huffnagle GB. The microbiome and the respiratory tract. Annu Rev Physiol 2016;78:481–504. https://doi.org/10.1146/annurev-physiol-021115-105238.

［4］Levine R, Evers C. The slow death of spontaneous generation (1668–1859). Biotech Chronicles. https://webprojects.oit.ncsu.edu/project/bio183de/Black/cellintro/cellintro_reading/Spontaneous_Generation.html [Accessed 5 July 2021].

［5］Jaynes ET. Probability theory: the logic of science. Cambridge: Cambridge University Press; 2003.

［6］Pearl J, Mackenzie D. The book of why: the new science of cause and effect. 1st. New York: Basic Books; 2018.

［7］Ding T, Schloss PD. Dynamics and associations of microbial community types across the human body. Nature 2014;509:357–360. https://doi.org/10.1038/nature13178.

[8] Valles-Colomer M, Falony G, Darzi Y, Tigchelaar EF, Wang J, Tito RY, et al. The neuroactive potential of the human gut microbiota in quality of life and depression. Nat Microbiol 2019. https://doi.org/10.1038/s41564-018-0337-x.

[9] Ben-Ari Y. Is birth a critical period in the pathogenesis of autism spectrum disorders? Nat Rev Neurosci 2015;16:498–505. https://doi.org/10.1038/nrn3956.

[10] Apps PJ, Weldon PJ, Kramer M. Chemical signals in terrestrial vertebrates: search for design features. Nat Prod Rep 2015;32:1131–1153. https://doi.org/10.1039/c5np00029g.

[11] Powell N, Walker MM, Talley NJ. The mucosal immune system: master regulator of bidirectional gut–brain communications. Nat Rev Gastroenterol Hepatol 2017;14:143–159. https://doi.org/10.1038/nrgastro.2016.191.

[12] Vujkovic-Cvijin I, Sklar J, Jiang L, Natarajan L, Knight R, Belkaid Y. Host variables confound gut microbiota studies of human disease. Nature 2020;2020:1–7. https://doi.org/10.1038/s41586-020-2881-9.

[13] Qi Q, Li J, Yu B, Moon J-Y, Chai JC, Merino J, et al. Host and gut microbial tryptophan metabolism and type 2 diabetes: an integrative analysis of host genetics, diet, gut microbiome and circulating metabolites in cohort studies. Gut 2021. https://doi.org/10.1136/gutjnl-2021-324053.

[14] Chassaing B, Koren O, Goodrich JK, Poole AC, Srinivasan S, Ley RE, et al. Dietary emulsifiers impact the mouse gut microbiota promoting colitis and metabolic syndrome. Nature 2015;519:92–96. https://doi.org/10.1038/nature14232.

[15] Turk JN, Zahavi ER, Gorman AE, Murray K, Turk MA, Veale DJ. Exploring the effect of alcohol on disease activity and outcomes in rheumatoid arthritis through systematic review and meta-analysis. Sci Rep 2021;11:10474. https://doi.org/10.1038/s41598-021-89618-1.

[16] Kolber MA, Walls RM, Hinners ML, Singer DS. Evidence of increased class I MHC expression on human peripheral blood lymphocytes during acute ethanol intoxication. Alcohol Clin Exp Res 1988;12:820–823. https://doi.org/10.1111/j.1530-0277.1988.tb01353.x.

[17] Nicolas GR, Chang PV. Deciphering the chemical lexicon of host–gut microbiota interactions. Trends Pharmacol Sci 2019;40:430–445. https://doi.org/10.1016/j.tips.2019.04.006.

[18] Azizov V, Dietel K, Steffen F, Dürholz K, Meidenbauer J, Lucas S, et al. Ethanol consumption inhibits TFH cell responses and the development of autoimmune arthritis. Nat Commun 2020;11:1998. https://doi.org/10.1038/s41467-020-15855-z.

[19] Wu W, Sun M, Chen F, Cao AT, Liu H, Zhao Y, et al. Microbiota metabolite short-chain fatty acid acetate promotes intestinal IgA response to microbiota which is mediated by GPR43. Mucosal Immunol 2017;10:946–956. https://doi.org/10.1038/mi.2016.114.

[20] Cao Y, Lin W, Li H. Large covariance estimation for compositional data via composition-adjusted thresholding. J Am Stat Assoc 2018;1–45. https://doi.org/10.1080/01621459.2018.1442340.

[21] Friedman J, Alm EJ. Inferring correlation networks from genomic survey data. PLoS Comput Biol 2012;8. https://doi.org/10.1371/journal.pcbi.1002687, e1002687.

[22] Holmes MV, Ala-Korpela M, Smith GD. Mendelian randomization in cardiometabolic disease: challenges in evaluating causality. Nat Rev Cardiol 2017;14:577–590. https://doi.org/10.1038/nrcardio.2017.78.

［23］Liu X, Tang S, Zhong H, Tong X, Jie Z, Ding Q, et al. A genome-wide association study for gut metagenome in Chinese adults illuminates complex diseases. Cell Discov 2021;7(1):9. https://doi.org/10.1038/s41421-020-00239-w.

［24］Jie Z, Xia H, Zhong S-L, Feng Q, Li S, Liang S, et al. The gut microbiome in atherosclerotic cardiovascular disease. Nat Commun 2017;8:845. https://doi.org/10.1038/s41467-017-00900-1.

［25］Liu X, Tong X, Zou Y, Lin X, Zhao H, Tian L, et al. Inter-determination of blood metabolite levels and gut microbiome supported by Mendelian randomization. BioRxiv 2020. https://doi.org/10.1101/2020.06.30.181438. 2020.06.30.

［26］Jiang X, Alfredsson L. Modifiable environmental exposure and risk of rheumatoid arthritis-current evidence from genetic studies. Arthritis Res Ther 2020;22. https://doi.org/10.1186/s13075-020-02253-5.

［27］Karlsson FH, Tremaroli V, Nookaew I, Bergström G, Behre CJ, Fagerberg B, et al. Gut metagenome in European women with normal, impaired and diabetic glucose control. Nature 2013;498:99–103. https://doi.org/10.1038/nature12198.

［28］Zhong H, Ren H, Lu Y, Fang C, Hou G, Yang Z, et al. Distinct gut metagenomics and metaproteomics signatures in prediabetics and treatment-naïve type 2 diabetics. EBioMedicine 2019. https://doi.org/10.1016/j.ebiom.2019.08.048.

［29］Jie Z, Liang S, Ding Q, Li F, Sun X, Lin Y, et al. Dairy consumption and physical fitness tests associated with fecal microbiome in a Chinese cohort. Med Microecol 2021. https://doi.org/10.1016/j.medmic.2021.100038.

［30］Schüssler-Fiorenza Rose SM, Contrepois K, Moneghetti KJ, Zhou W, Mishra T, Mataraso S, et al. A longitudinal big data approach for precision health. Nat Med 2019;25:792–804. https://doi.org/10.1038/s41591-019-0414-6.

［31］Qin J, Li Y, Cai Z, Li S, Zhu J, Zhang F, et al. A metagenome-wide association study of gut microbiota in type 2 diabetes. Nature 2012;490:55–60. https://doi.org/10.1038/nature11450.

［32］Jackson M, Jeffery IB, Beaumont M, Bell JT, Clark AG, Ley RE, et al. Signatures of early frailty in the gut microbiota. Genome Med 2016;8:8. https://doi.org/10.1186/s13073-016-0262-7.

［33］Jie Z, Liang S, Ding Q, Li F, Tang S, Wang D, et al. A transomic cohort as a reference point for promoting a healthy gut microbiome. Med Microecol 2021. https://doi.org/10.1016/j.medmic.2021.100039.

［34］Png CW, Lindén SK, Gilshenan KS, Zoetendal EG, McSweeney CS, Sly LI, et al. Mucolytic bacteria with increased prevalence in IBD mucosa augment in vitro utilization of mucin by other bacteria. Am J Gastroenterol 2010;105:2420–2428. https://doi.org/10.1038/ajg.2010.281.

［35］Liu X, Tong X, Zhu J, Tian L, Jie Z, Zou Y, et al. Metagenome-genome-wide association studies reveal human genetic impact on the oral microbiome. bioRxiv 2021. https://doi.org/10.1101/2021.05.06.443017.

［36］Zhang Y, Gu Y, Ren H, Wang S, Zhong H, Zhao X, et al. Gut microbiome-related effects of berberine and probiotics on type 2 diabetes (the PREMOTE study). Nat Commun 2020;11:5015. https://doi.org/10.1038/s41467-020-18414-8.

［37］Cohen CR, Wierzbicki MR, French AL, Morris S, Newmann S, Reno H, et al. Randomized trial of lactin-V to prevent recurrence of bacterial vaginosis. N Engl J Med 2020;382:1906–1915.

https://doi.org/10.1056/NEJMoa1915254.

[38] Xu F, Fu Y, Sun TY, Jiang Z, Miao Z, Shuai M, et al. The interplay between host genetics and the gut microbiome reveals common and distinct microbiome features for complex human diseases. Microbiome 2020;8:145. https://doi.org/10.1186/s40168-020-00923-9.

[39] Pedersen HK, Gudmundsdottir V, Nielsen HB, Hyotylainen T, Nielsen T, Jensen BAH, et al. Human gut microbes impact host serum metabolome and insulin sensitivity. Nature 2016;535:376–381. https://doi.org/10.1038/nature18646.

[40] Mukherji A, Kobiita A, Ye T, Chambon P. Homeostasis in intestinal epithelium is orchestrated by the circadian clock and microbiota cues transduced by TLRs. Cell 2013;153:812–827. https://doi.org/10.1016/j.cell.2013.04.020.

[41] Braniste V, Al-Asmakh M, Kowal C, Anuar F, Abbaspour A, Tóth M, et al. The gut microbiota influences blood-brain barrier permeability in mice. Sci Transl Med 2014;6. https://doi.org/10.1126/scitranslmed.3009759, 263ra158.

[42] Olszak T, An D, Zeissig S, Vera MP, Richter J, Franke A, et al. Microbial exposure during early life has persistent effects on natural killer T cell function. Science 2012;336:489–493. https://doi.org/10.1126/science.1219328.

[43] Xiao L, Feng Q, Liang S, Sonne SB, Xia Z, Qiu X, et al. A catalog of the mouse gut metagenome. Nat Biotechnol 2015;33:1103–1108. https://doi.org/10.1038/nbt.3353.

[44] Rausch P, Basic M, Batra A, Bischoff SC, Blaut M, Clavel T, et al. Analysis of factors contributing to variation in the C57BL/6J fecal microbiota across German animal facilities. Int J Med Microbiol 2016. https://doi.org/10.1016/j.ijmm.2016.03.004.

[45] Wroblewski LE, Peek RM, Wilson KT. Helicobacter pylori and gastric cancer: factors that modulate disease risk. Clin Microbiol Rev 2010;23:713–739. https://doi.org/10.1128/CMR.00011-10.

[46] Chen XH, Wang A, Chu AN, Gong YH, Yuan Y. Mucosa-associated microbiota in gastric cancer tissues compared with non-cancer tissues. Front Microbiol 2019;10:1261. https://doi.org/10.3389/fmicb.2019.01261.

[47] Bugaytsova JA, Björnham O, Chernov YA, Gideonsson P, Henriksson S, Mendez M, et al. Helicobacter pylori adapts to chronic infection and gastric disease via pH-responsive BabA-mediated adherence. Cell Host Microbe 2017;21:376–389. https://doi.org/10.1016/j.chom.2017.02.013.

[48] Poole AC, Goodrich JK, Youngblut ND, Luque GG, Ruaud A, Sutter JL, et al. Human salivary amylase gene copy number impacts oral and gut microbiomes. Cell Host Microbe 2019;25(4):553–564.e7. https://doi.org/10.1016/j.chom.2019.03.001.

[49] Zhu J, Tian L, Chen P, Han M, Song L, Tong X, et al. Over 50000 metagenomically assembled draft genomes for the human oral microbiome reveal new taxa. Genomics Proteomics Bioinformatics 2021;18:2790. https://doi.org/10.1016/j.gpb.2021.05.001.

[50] Atarashi K, Suda W, Luo C, Kawaguchi T, Motoo I, Narushima S, et al. Ectopic colonization of oral bacteria in the intestine drives T H 1 cell induction and inflammation. Science 2017;358:359–365. https://doi.org/10.1126/science.aan4526.

[51] Bajaj JS, Matin P, White MB, Fagan A, Golob Deeb J, Acharya C, et al. Periodontal therapy

favorably modulates the oral-gut-hepatic axis in cirrhosis. Am J Physiol Liver Physiol 2018. https://doi.org/10.1152/ajpgi.00230.2018. ajpgi.00230.2018.

[52] Sobko T, Reinders CI, Jansson E, Norin E, Midtvedt T, Lundberg JO. Gastrointestinal bacteria generate nitric oxide from nitrate and nitrite. Nitric Oxide 2005;13:272–278. https://doi.org/10.1016/j.niox.2005.08.002.

[53] Hata DJ, Smith DS. Blood group B degrading activity of Ruminococcus gnavus α-galactosidase. Artif Cells Blood Substit Immobil Biotechnol 2004;32:263–274. https://doi.org/10.1081/BIO-120037831.

[54] Tailford LE, Owen CD, Walshaw J, Crost EH, Hardy-Goddard J, Le Gall G, et al. Discovery of intramolecular trans-sialidases in human gut microbiota suggests novel mechanisms of mucosal adaptation. Nat Commun 2015;6:7624. https://doi.org/10.1038/ncomms8624.

[55] Henke MT, Kenny DJ, Cassilly CD, Vlamakis H, Xavier RJ, Clardy J. Ruminococcus gnavus, a member of the human gut microbiome associated with Crohn's disease, produces an inflammatory polysaccharide. Proc Natl Acad Sci U S A 2019;116:12672–12677. https://doi.org/10.1073/pnas.1904099116.

[56] Ng KM, Ferreyra JA, Higginbottom SK, Lynch JB, Kashyap PC, Gopinath S, et al. Microbiota-liberated host sugars facilitate post-antibiotic expansion of enteric pathogens. Nature 2013;502:96–99. https://doi.org/10.1038/nature12503.

[57] McGregor JA, French JI, Jones W, Milligan K, McKinney PJ, Patterson E, et al. Bacterial vaginosis is associated with prematurity and vaginal fluid mucinase and sialidase: results of a controlled trial of topical clindamycin cream. Am J Obstet Gynecol 1994;170:1048–1059. discussion 1059-60 https://doi.org/10.1016/s0002-9378(94)70098-2.

[58] Dos Santos Santiago GL, Tency I, Verstraelen H, Verhelst R, Trog M, Temmerman M, et al. Longitudinal qPCR study of the dynamics of L. crispatus, L. iners, A. vaginae, (sialidase positive) G. vaginalis, and P. bivia in the vagina. PLoS One 2012;7. https://doi.org/10.1371/journal.pone.0045281, e45281.

[59] Bacher P, Hohnstein T, Beerbaum E, Röcker M, Blango MG, Kaufmann S, et al. Human anti-fungal Th17 immunity and pathology rely on cross-reactivity against Candida albicans. Cell 2019. https://doi.org/10.1016/j.cell.2019.01.041.

[60] Boldock E, Surewaard BGJ, Shamarina D, Na M, Fei Y, Ali A, et al. Human skin commensals augment Staphylococcus aureus pathogenesis. Nat Microbiol 2018. https://doi.org/10.1038/s41564-018-0198-3.

[61] Everard A, Belzer C, Geurts L, Ouwerkerk JP, Druart C, Bindels LB, et al. Cross talk between Akkermansia muciniphila and intestinal epithelium controls diet-induced obesity. Proc Natl Acad Sci U S A 2013;110:9066–9071. https://doi.org/10.1073/pnas.1219451110.

[62] Seck EH, Senghor B, Merhej V, Bachar D, Cadoret F, Robert C, et al. Salt in stools is associated with obesity, gut halophilic microbiota and Akkermansia muciniphila depletion in humans. Int J Obes (Lond) 2019;43:862–871. https://doi.org/10.1038/s41366-018-0201-3.

[63] Liou AP, Paziuk M, Luevano JM, Machineni S, Turnbaugh PJ, Kaplan LM. Conserved shifts in the gut microbiota due to gastric bypass reduce host weight and adiposity. Sci Transl Med 2013;5:178ra41. https://doi.org/10.1126/scitranslmed.3005687.

［64］Wu H, Esteve E, Tremaroli V, Khan MT, Caesar R, Mannerås-Holm L, et al. Metformin alters the gut microbiome of individuals with treatment-naive type 2 diabetes, contributing to the therapeutic effects of the drug. Nat Med 2017;23:850–858. https://doi.org/10.1038/nm.4345.

［65］Plovier H, Everard A, Druart C, Depommier C, Van Hul M, Geurts L, et al. A purified membrane protein from Akkermansia muciniphila or the pasteurized bacterium improves metabolism in obese and diabetic mice. Nat Med 2017;23:107–113. https://doi.org/10.1038/nm.4236.

［66］Depommier C, Everard A, Druart C, Plovier H, Van Hul M, Vieira-Silva S, et al. Supplementation with Akkermansia muciniphila in overweight and obese human volunteers: a proof-of-concept exploratory study. Nat Med 2019. https://doi.org/10.1038/s41591-019-0495-2.

［67］Chevalier C, Stojanović O, Colin DJ, Suarez-Zamorano N, Tarallo V, Veyrat-Durebex C, et al. Gut microbiota orchestrates energy homeostasis during cold. Cell 2015;163:1360–1374. https://doi.org/10.1016/j.cell.2015.11.004.

［68］Yoon HS, Cho CH, Yun MS, Jang SJ, You HJ, Hyeong KJ, et al. Akkermansia muciniphila secretes a glucagon-like peptide-1-inducing protein that improves glucose homeostasis and ameliorates metabolic disease in mice. Nat Microbiol 2021;1–11. https://doi.org/10.1038/s41564-021-00880-5.

［69］Perry RJ, Peng L, Barry NA, Cline GW, Zhang D, Cardone RL, et al. Acetate mediates a microbiome–brain–β-cell axis to promote metabolic syndrome. Nature 2016;534:213–217. https://doi.org/10.1038/nature18309.

［70］Sanna S, van Zuydam NR, Mahajan A, Kurilshikov A, Vich Vila A, Võsa U, et al. Causal relationships among the gut microbiome, short-chain fatty acids and metabolic diseases. Nat Genet 2019;51:600–605. https://doi.org/10.1038/s41588-019-0350-x.

［71］De Vadder F, Kovatcheva-Datchary P, Zitoun C, Duchampt A, Bäckhed F, Mithieux G. Microbiota-produced succinate improves glucose homeostasis via intestinal gluconeogenesis. Cell Metab 2016;24:151–157. https://doi.org/10.1016/j.cmet.2016.06.013.

［72］Wen C, Yan W, Mai C, Duan Z, Zheng J, Sun C, et al. Joint contributions of the gut microbiota and host genetics to feed efficiency in chickens. Microbiome 2021;9:126. https://doi.org/10.1186/s40168-021-01040-x.

［73］Weir TL, Manter DK, Sheflin AM, Barnett BA, Heuberger AL, Ryan EP. Stool microbiome and metabolome differences between colorectal cancer patients and healthy adults. PLoS One 2013;8. https://doi.org/10.1371/journal.pone.0070803, e70803.

［74］Feng Q, Liang S, Jia H, Stadlmayr A, Tang L, Lan Z, et al. Gut microbiome development along the colorectal adenoma–carcinoma sequence. Nat Commun 2015;6:6528. https://doi.org/10.1038/ncomms7528.

［75］Liu P, Wu L, Peng G, Han Y, Tang R, Ge J, et al. Altered microbiomes distinguish Alzheimer's disease from amnestic mild cognitive impairment and health in a Chinese cohort. Brain Behav Immun 2019. https://doi.org/10.1016/j.bbi.2019.05.008.

［76］Zhu F, Ju Y, Wang W, Wang Q, Guo R, Ma Q, et al. Metagenome-wide association of gut microbiome features for schizophrenia. Nat Commun 2020;11:1612. https://doi.org/10.1038/s41467-020-15457-9.

［77］Xie H, Guo R, Zhong H, Feng Q, Lan Z, Qin B, et al. Shotgun metagenomics of 250 adult twins reveals genetic and environmental impacts on the gut microbiome. Cell Syst 2016;3:572–584.e3. https://doi.org/10.1016/j.cels.2016.10.004.

[78] Guo X, Li S, Zhang J, Wu F, Li X, Wu D, et al. Genome sequencing of 39 Akkermansia muciniphila isolates reveals its population structure, genomic and functional diveriristy, and global distribution in mammalian gut microbiotas. BMC Genomics 2017;18:800. https://doi.org/10.1186/s12864-017-4195-3.

[79] Johansson MEV, Hansson GC. Immunological aspects of intestinal mucus and mucins. Nat Rev Immunol 2016;16:639–649. https://doi.org/10.1038/nri.2016.88.

[80] Ansaldo E, Slayden LC, Ching KL, Koch MA, Wolf NK, Plichta DR, et al. Akkermansia muciniphila induces intestinal adaptive immune responses during homeostasis. Science 2019;364:1179–1184. https://doi.org/10.1126/science.aaw7479.

[81] Tett A, Huang KD, Asnicar F, Fehlner-Peach H, Pasolli E, Karcher N, et al. The Prevotella copri complex comprises four distinct clades underrepresented in westernized populations. Cell Host Microbe 2019. https://doi.org/10.1016/j.chom.2019.08.018.

[82] Newgard CB. Interplay between lipids and branched-chain amino acids in development of insulin resistance. Cell Metab 2012;15:606–614. https://doi.org/10.1016/j.cmet.2012.01.024.

[83] Neinast MD, Jang C, Hui S, Murashige DS, Chu Q, Morscher RJ, et al. Quantitative analysis of the whole-body metabolic fate of branched-chain amino acids. Cell Metab 2019;29:417–429.e4. https://doi.org/10.1016/j.cmet.2018.10.013.

[84] Lim MT, Pan BJ, Toh DWK, Sutanto CN, Kim JE. Animal protein versus plant protein in supporting lean mass and muscle strength: a systematic review and meta-analysis of randomized controlled trials. Nutrients 2021;13. https://doi.org/10.3390/nu13020661.

[85] Duan Y, Guo Q, Wen C, Wang W, Li Y, Tan B, et al. Free amino acid profile and expression of genes implicated in protein metabolism in skeletal muscle of growing pigs fed low-protein diets supplemented with branched-chain amino acids. J Agric Food Chem 2016;64:9390–9400. https://doi.org/10.1021/acs.jafc.6b03966.

[86] Yu D, Richardson NE, Green CL, Spicer AB, Murphy ME, Flores V, et al. The adverse metabolic effects of branched-chain amino acids are mediated by isoleucine and valine. Cell Metab 2021;33:905–922.e6. https://doi.org/10.1016/j.cmet.2021.03.025.

[87] Ikeda K, Kinoshita M, Kayama H, Nagamori S, Kongpracha P, Umemoto E, et al. Slc3a2 mediates branched-chain amino-acid-dependent maintenance of regulatory T cells. Cell Rep 2017;21:1824–1838. https://doi.org/10.1016/j.celrep.2017.10.082.

[88] He Q, Gao Y, Jie Z, Yu X, Laursen JMJM, Xiao L, et al. Two distinct metacommunities characterize the gut microbiota in Crohn's disease patients. Gigascience 2017;6:1–11. https://doi.org/10.1093/gigascience/gix050.

[89] De Groot P, Nikolic T, Pellegrini S, Sordi V, Imangaliyev S, Rampanelli E, et al. Faecal microbiota transplantation halts progression of human new-onset type 1 diabetes in a randomised controlled trial. Gut 2020. https://doi.org/10.1136/gutjnl-2020-322630. gutjnl-2020-322630.

[90] Paun A, Yau C, Meshkibaf S, Daigneault MC, Marandi L, Mortin-Toth S, et al. Association of HLA-dependent islet autoimmunity with systemic antibody responses to intestinal commensal bacteria in children. Sci Immunol 2019;4. https://doi.org/10.1126/sciimmunol.aau8125,

eaau8125.

[91] Vatanen T, Kostic AD, D'Hennezel E, Siljander H, Franzosa EA, Yassour M, et al. Variation in microbiome LPS immunogenicity contributes to autoimmunity in humans. Cell 2016;165:842–853. https://doi.org/10.1016/j.cell.2016.04.007.

[92] Hebbandi Nanjundappa R, Ronchi F, Wang J, Clemente-Casares X, Yamanouchi J, Sokke Umeshappa C, et al. A gut microbial mimic that hijacks diabetogenic autoreactivity to suppress colitis. Cell 2017;171:655–667.e17. https://doi.org/10.1016/j.cell.2017.09.022.

[93] Wegorzewska MM, Glowacki RWP, Hsieh SA, Donermeyer DL, Hickey CA, Horvath SC, et al. Diet modulates colonic T cell responses by regulating the expression of a Bacteroides thetaiotaomicron antigen. Sci Immunol 2019;4:eaau9079. https://doi.org/10.1126/sciimmunol.aau9079.

[94] Zhang X, Zhang D, Jia H, Feng Q, Wang D, Di Liang D, et al. The oral and gut microbiomes are perturbed in rheumatoid arthritis and partly normalized after treatment. Nat Med 2015;21:895–905. https://doi.org/10.1038/nm.3914.

[95] Zhou C, Zhao H, Xiao X, Chen B, Guo R, Wang Q, et al. Metagenomic profiling of the pro-inflammatory gut microbiota in ankylosing spondylitis. J Autoimmun 2020;107:102360. https://doi.org/10.1016/j.jaut.2019.102360.

[96] Szymula A, Rosenthal J, Szczerba BM, Bagavant H, Fu SM, Deshmukh US. T cell epitope mimicry between Sjögren's syndrome antigen a (SSA)/Ro60 and oral, gut, skin and vaginal bacteria. Clin Immunol 2014;152:1–9. https://doi.org/10.1016/j.clim.2014.02.004.

[97] Greiling TM, Dehner C, Chen X, Hughes K, Iñiguez AJ, Boccitto M, et al. Commensal orthologs of the human autoantigen Ro60 as triggers of autoimmunity in lupus. Sci Transl Med 2018;10. https://doi.org/10.1126/scitranslmed.aan2306, eaan2306.

[98] Agus A, Planchais J, Sokol H. Gut microbiota regulation of tryptophan metabolism in health and disease. Cell Host Microbe 2018;23:716–724. https://doi.org/10.1016/j.chom.2018.05.003.

[99] Geller LT, Barzily-Rokni M, Danino T, Jonas OH, Shental N, Nejman D, et al. Potential role of intratumor bacteria in mediating tumor resistance to the chemotherapeutic drug gemcitabine. Science 2017;357:1156–1160. https://doi.org/10.1126/science.aah5043.

[100] Yang D, Chen X, Wang J, Lou Q, Lou Y, Li L, et al. Dysregulated lung commensal bacteria drive interleukin-17B production to promote pulmonary fibrosis through their outer membrane vesicles. Immunity 2019;50:692–706.e7. https://doi.org/10.1016/j.immuni.2019.02.001.

[101] Toyofuku M, Nomura N, Eberl L. Types and origins of bacterial membrane vesicles. Nat Rev Microbiol 2019;17:13–24. https://doi.org/10.1038/s41579-018-0112-2.

[102] Ñahui Palomino RA, Vanpouille C, Laghi L, Parolin C, Melikov K, Backlund P, et al. Extracellular vesicles from symbiotic vaginal lactobacilli inhibit HIV-1 infection of human tissues. Nat Commun 2019;10:5656. https://doi.org/10.1038/s41467-019-13468-9.

[103] Duan Y, Llorente C, Lang S, Brandl K, Chu H, Jiang L, et al. Bacteriophage targeting of gut bacterium attenuates alcoholic liver disease. Nature 2019. https://doi.org/10.1038/s41586-019-1742-x.

第7章

宏基因组的临床应用

摘　要：前面的章节已经为我们提供了关于人类菌群研究的理论和实践知识。本章将讨论如何将这些知识应用于临床。生物标志物，包括各种疾病的病原体（第6章），使得开展人群规模的筛查变得可行，持续进行临床研究，并完善用于诊断和治疗的微生物组模型也很重要。医疗保健专业人员需要决定是否将手术中的样本用于研究，是否在诸如癌症免疫治疗及生殖治疗前后进行宏基因组学检测。本章可协助来自世界各地不同专业背景的医生思考宏基因组学如何帮助患者，有助于推动技术和临床需求标准的进一步发展。

关键词：癌症筛查，生物标志物，癌症治疗，肝病，精准医学，药物代谢，临床宏基因组学

7.1　疾病筛查领域的宏基因组研究

根据前几章可知，位于身体不同部位的菌群可能反映并且贡献了过去几十年人类疾病的变化趋势（图7.1）。当我们试图在一个欠发达地区改善生活条件、营养状况和提高生活质量时，需要意识到并预期正常菌群和疾病流行程度将会发生的变化。

虽然科学家们总是对宏基因组测序技术充满兴趣，但是清楚地了解一种疾病的真正需求，对于在医院常规地使用宏基因组测序非常重要（表7.1）。比如一个即将昏迷的肝病患者是否需要宏基因组检测？将要接受骨髓移植的白血病患儿是否需要宏基因组检测？成人的数据显示，移植前和移植期间粪便微生物组多样性非常低可能会夺去一个人的生命[1-3]；类似的证据也出现在儿科患者身上，包括骨髓移植前后的粪便、口腔和鼻腔微生物组检测中[4,5]。

随着乙型肝炎疫苗的接种、黄曲霉素暴露量的降低以及治疗丙型肝炎的药物的使用，肝病的发展趋势已经发生了转变。随着全球肥胖人数的增加，非酒精性脂肪性肝病（non-alcoholic fatty liver disease，NAFLD）是目前最流行的会进展为急性住院治疗或转为肝细胞癌（hepatocellular carcinoma，HCC）的肝脏疾病。NAFLD包括肝脂肪变性（脂肪肝）、非酒精性脂肪性肝炎（non-alcoholic steatohepatitis，NASH）和肝硬化[6]。脂肪肝在超声检查中十分常见。

从常规体检中，可看出肝酶丙氨酸氨基转移酶和 γ- 谷氨酰转肽酶水平偏高 [7]。NAFLD 患者有可能发展为肝硬化，再从肝硬化发展为肝癌，粪便和口腔微生物组有助于预测哪些患者更有可能经历这一历程。男性肝癌的高发也可以从肠道微生物中产生的次级胆汁酸水平的性别差异找到解释 [8]。非酒精性脂肪肝的可能机制包括某些肺炎克雷伯菌菌株产生乙醇 [9]。微生物组在急性症状中可能同样重要。治疗疼痛和感冒常用的对乙酰氨基酚是导致急性肝衰竭的主要原因，夜间用药则会导致更严重的肝损伤。在一项小鼠研究中发现肠道微生物代谢物，如 1- 苯基 -1,2- 丙二酮，会消耗肝脏的谷胱甘肽，这能解释对乙酰氨基酚毒性的昼夜差异 [10]。

Females

Leading causes 1990	Leading causes 2007	Mean percentage change in number of prevalent cases, 1990–2007	Mean percentage change in all-age prevalence rate, 1990–2007	Mean percentage change in age-standardised prevalence rate, 1990–2007	Leading causes 2017	Mean percentage change in number of prevalent cases, 2007–17	Mean percentage change in all-age prevalence rate, 2007–17	Mean percentage change in age-standardised prevalence rate, 2007–17
1 Oral disorders	1 Oral disorders	23·1	-2·0	-3·8	1 Oral disorders	13·5	0·3	-1·3
2 Headache disorders	2 Headache disorders	31·5	4·7	-0·4	2 Headache disorders	14·5	1·2	0·3
3 Haemoglobinopathies	3 Haemoglobinopathies	29·9	3·3	4·2	3 Haemoglobinopathies	13·4	0·2	0·8
4 Tuberculosis	4 Tuberculosis	27·7	1·6	-2·2	4 Tuberculosis	1·2	-10·6	-11·7
5 Intestinal nematode	5 Gynaecological diseases	34·0	6·6	-2·3	5 Gynaecological diseases	13·3	0·1	-0·5
6 Dietary iron deficiency	6 STIs	40·2	11·6	1·7	6 STIs	17·7	4·0	0·7
7 Gynaecological diseases	7 Dietary iron deficiency	7·2	-14·7	-14·5	7 Blindness and vision impairment	24·1	9·7	0·7
8 STIs	8 Blindness and vision impairment	43·4	14·1	0·9	8 Age-related hearing loss	26·1	11·4	0·9
9 Blindness and vision impairment	9 Intestinal nematode	-20·7	-36·9	-34·9	9 Dietary iron deficiency	6·4	-6·0	-4·9
10 Cirrhosis	10 Age-related hearing loss	45·4	15·7	1·2	10 Cirrhosis	23·5	9·2	4·6
11 Age-related hearing loss	11 Cirrhosis	40·8	12·0	5·0	11 Intestinal nematode	-15·7	-25·5	-23·4
12 Vitamin A deficiency	12 Vitamin A deficiency	11·4	-11·3	-5·2	12 Upper digestive diseases	21·1	7·0	1·5
13 Fungal skin diseases	13 Upper digestive diseases	37·1	9·1	-1·2	13 Chronic kidney disease	28·2	13·3	3·0
14 Upper digestive diseases	14 Fungal skin diseases	23·0	-2·1	-3·0	14 Vitamin A deficiency	5·9	-6·4	-4·0
15 Chronic kidney disease	15 Chronic kidney disease	43·2	14·0	-1·3	15 Fungal skin diseases	12·5	-0·6	-4·0
16 Low back pain	16 Low back pain	29·6	3·2	-7·7	16 Low back pain	17·4	3·8	-2·7
17 Other skin diseases	17 Other skin diseases	44·2	14·8	5·7	17 Other skin diseases	25·4	10·8	3·9
18 Interpersonal violence	18 Diabetes	70·2	35·4	17·6	18 Diabetes	29·8	14·7	3·8
19 Iodine deficiency	19 Interpersonal violence	28·1	1·9	-2·3	19 Interpersonal violence	14·7	1·4	1·1
20 Anxiety disorders	20 Anxiety disorders	33·1	5·9	0·3	20 Other musculoskeletal	21·6	7·5	0·9
26 Diabetes	21 Other musculoskeletal				23 Anxiety disorders			
27 Other musculoskeletal	33 Iodine deficiency				35 Iodine deficiency			

Males

Leading causes 1990	Leading causes 2007	Mean percentage change in number of prevalent cases, 1990–2007	Mean percentage change in all-age prevalence rate, 1990–2007	Mean percentage change in age-standardised prevalence rate, 1990–2007	Leading causes 2017	Mean percentage change in number of prevalent cases, 2007–17	Mean percentage change in all-age prevalence rate, 2007–17	Mean percentage change in age-standardised prevalence rate, 2007–17
1 Oral disorders	1 Oral disorders	21·6	-2·9	-4·3	1 Oral disorders	12·5	-0·2	-1·6
2 Headache disorders	2 Headache disorders	31·3	4·8	0·0	2 Headache disorders	14·3	1·5	0·7
3 Tuberculosis	3 Tuberculosis	26·2	0·7	-3·1	3 Tuberculosis	1·1	-10·2	-11·5
4 Intestinal nematode	4 Cirrhosis	42·5	13·8	6·5	4 Cirrhosis	22·8	9·0	4·6
5 Cirrhosis	5 Haemoglobinopathies	29·0	3·0	3·6	5 Age-related hearing loss	24·3	10·3	0·0
6 Dietary iron deficiency	6 Intestinal nematode	-21·4	-37·3	-35·7	6 Haemoglobinopathies	12·7	0·1	0·7
7 Haemoglobinopathies	7 Age-related hearing loss	44·6	15·4	0·4	7 Blindness and vision impairment	23·1	9·3	-0·4
8 Age-related hearing loss	8 Dietary iron deficiency	6·0	-15·4	-14·3	8 Dietary iron deficiency	5·8	-6·1	-5·2
9 Vitamin A deficiency	9 Blindness and vision impairment	39·5	11·3	-2·2	9 STIs	19·7	6·3	1·9
10 Blindness and vision impairment	10 Vitamin A deficiency	9·7	-12·4	-7·1	10 Intestinal nematode	-16·7	-26·0	-24·2
11 Fungal skin diseases	11 STIs	38·9	10·9	0·7	11 Vitamin A deficiency	5·6	-6·3	-4·0
12 STIs	12 Fungal skin diseases	20·8	-3·5	-3·5	12 Upper digestive diseases	20·3	6·8	1·3
13 Upper digestive diseases	13 Upper digestive diseases	36·5	9·0	-1·3	13 Fungal skin diseases	10·2	-2·2	-4·6
14 Low back pain	14 Chronic kidney disease	45·6	16·2	-0·1	14 Chronic kidney disease	25·4	11·4	1·1
15 Chronic kidney disease	15 Low back pain	30·3	4·0	-6·8	15 Other skin diseases	26·4	12·2	4·7
16 Other skin diseases	16 Other skin diseases	46·5	16·9	7·1	16 Low back pain	18·0	4·7	-1·3
17 Falls	17 Diabetes	77·6	41·8	21·5	17 Diabetes	29·3	14·8	4·0
18 Diabetes	18 Falls	26·4	0·9	-9·8	18 Falls	26·8	12·6	4·1
19 Asthma	19 Other musculoskeletal	41·6	13·0	0·8	19 Other musculoskeletal	16·7	3·6	-2·9
20 Dermatitis	20 COPD	31·5	4·9	-10·6	20 COPD	15·6	2·6	-10·1
21 COPD	21 Dermatitis				22 Dermatitis			
22 Other musculoskeletal	22 Asthma				23 Asthma			

☐ Communicable, maternal, neonatal, and nutritional diseases
☐ Non-communicable diseases
☐ Injuries

图 7.1　不同身体部位的菌群反映的人类疾病的变化趋势

1990年、2007年和2017年致病率最高的20个具有三级证据级别的疾病致病因素，每个性别的病例数量以及各年龄和年龄标准化比率的百分比变化。第一级包括三大类别的病因：传染性疾病；孕产妇疾病、新生儿疾病和营养性疾病；非传染性疾病和伤害。对于非致命的健康评价，有22个二级原因，167个三级原因，288个四级原因。原因通过时间段之间的线段连接起来；实线代表增加，虚线代表减少。在1990—2007年和2007—2017年期间，显示了三种变化措施：病例数量的百分比变化、各年龄段流行率的百分比变化以及按年龄标准化后流行率的百分比变化。传染病、产妇疾病、新生儿疾病和营养疾病用红色显示，非传染病原因用蓝色显示，伤害用绿色显示。统计上显著性差异以粗体显示。COPD：慢性阻塞性肺病；STI：性传播疾病。来源：Lancet, 2018, 392:1789-1858. https://doi.org/10.1016/S0140-6736（18）32279-7.

表 7.1 临床实践中微生物检测的考虑事项

考虑事项	可用技术
结果需要多快提供？	
结果需要达到多高的敏感性和准确性？	
是否需要结合其他技术进行检测？	
目前可选用作为微生物组检测补充的黄金标准实践是什么（用于减少假阴性结果，并确认阳性诊断）？	是否需要专家小组的评估？
检测费用是多少，由谁承担？	

结肠镜检查是现今许多发达国家的常规体检项目。相对于粪便隐血试验和宿主基因甲基化的定量 PCR，宏基因组测序技术或者对少数特定菌进行 qPCR 可能是一种更方便、更可靠的检测技术。从政府的角度来看可以节省部分患者结肠镜检查费用；从个人角度可减少肠镜前洗肠引起的微生物群紊乱，而对于宏基因组测序病例的患者来说，则可能知道自己患其他疾病的风险和治疗方案的选择（详见第 8 章）。

对于肺癌，对痰液样本进行宏基因组学检测可能有助于缩短医院内 CT 的等待时间，并可能与细胞游离 DNA（cfDNA）（示例 7.1）检测相结合，提高两种技术在筛查潜在患者和防止复发方面的能力。口腔微生物组也被报道具有胰腺癌等疾病的生物标志物[12-14]，有待进一步的验证。根据孟德尔随机化分析[15]的结果（示例 6.2），粪便反硝化无色杆菌似乎促进了胆管癌。更广泛地说，微生物群落可能是世界各地不同人群中不同疾病发病率的关键因素（图 8.1，第 8 章）。

示例 7.1　人体体液中的游离 DNA 或 RNA

在许多国家，使用孕妇的 cfDNA 进行的产前疾病筛查已经成为守护分娩的产前诊断。利用血浆样本中的 cfDNA 或无细胞 RNA（cfRNA）进行肿瘤筛查和复发预测，也逐渐进入临床[18-20]。虽然覆盖率通常很低，而且数据可能本身就很零碎[21]，但是这种血浆 cfDNA 或 cfRNA 中的非人源读段可以定位到病毒、细菌和真菌。对于所有这些微生物，快速测序和生物信息学分析可以让医生知道各分类群在患者中的丰富程度，连同耐药基因、毒性和由宏基因组衍生的微生物基因组谱系信息（第 5 章），远快于尝试体外培养所有可能的微生物。在早期的应用中，病原体数据库和人群基线并不完善。

在疑似绒毛膜羊膜炎感染（包围胎儿和羊水的细胞膜感染）病例的脐带血中的 cfDNA 与健康对照组相比，可检测到富集的细菌[22]。对于侵入性真菌感染，例如，在器官移植后的慢性免疫抑制期间，血浆 cfDNA 鉴定的真菌，9 个患者中有 7 个患者的结果与平板培养实验或靶向测序结果吻合良好，可以减少组织活检的需要[23]。宏基因组学的优势更多地在于对微生物的无偏检测，即使是在培养阴性或（针对性）PCR 阴性的样本中（第 1 章，图 1.1、图 1.2）[16,17]，宏基因组检测将成为第一个看到同样症状的病原体变化趋势的检测。

对于心血管疾病（第 4 章）和精神疾病，口腔或粪便宏基因组样本可以定期从患者家中送检，以捕获复发的先兆，这也可能是检测癌症患者复发早期征兆的有效方法。对于人类基因序列比例较高的体液样本，游离 DNA 或 RNA 也是检测病原体的一种选择（示例 7.1）[16,17]，虽然这会失去了细胞内和附着的微生物。

为了进行诊断，必须注意方法和模型的假阴性和假阳性（表 7.1）。危及生命的疾病需要尽早诊断，因此阈值通常不是使 AUC 最大的值，而是倾向于使假阴性最小的值（例如，10 000 人中有 1 人漏诊可以被容许吗？）。以防这种罕见事件发生，可用保险与检测相结合。假阳性可以用另一种现有的检测方法进行验证，但是人们还需要估计在当前使用的阈值下，所预期的进一步检测的次数。

在美国，在一项多中心使用血液检测（人类游离 DNA 和蛋白质生物标志物）的癌症早筛临床试验中，对 10 006 名年龄 65 ~ 75 岁的妇女进行检查，之后进行 PET-CT 成像发现了 26 例癌症病例，而传统方法检查出了另外 24 例[18]。实际上，那些检测阳性但是没有出现临床症状的患者仍然应该在随后的几年中被随访。因此，这些测试结果并不一定是假阳性，而是反映了临床表现在病程和严重程度上的个体差异。从伦理和经济上讲，每个全民筛查都应该精心设计，以尽量减少不必要的焦虑和成本。大多数筛查是在老年人身上进行的，因为年轻人的发病率太低，而且测试会导致过多的假阳性（大量健康个体中的一小部分仍然是一个很大的数字）。不过，年轻人可能会更有兴趣参与没有特定研究问题的纵向研究（第 8 章）。

表 7.2 微生物和对癌症治疗的响应

癌症	治疗方案	微生物	参考文献
鼠皮下注射纤维肉瘤、黑色素瘤或肥大细胞瘤细胞系	CpG-ODN 免疫治疗；铂类化疗药物（奥沙利铂）	对 CpG-ODN 治疗，另枝菌 Alistipes shahii、瘤胃球菌与肿瘤内肿瘤坏死因子（TNF）表达呈正相关；乳杆菌与 TNF 表达呈负相关；动物实验证一步验证了另枝菌（A. shahii）和发酵乳杆菌（L. fermentum）的作用	[24]
小鼠皮下注射黑色素瘤、淋巴瘤或结肠癌细胞系	环磷酰胺（CTX）	环磷酰胺诱导乳杆菌（如约翰逊乳杆菌）和海氏肠球菌向次级淋巴器官移位。这些革兰氏阳性菌是诱导 Th17 细胞介导 CTX 效应所必需的	[25]
小鼠皮下注射纤维肉瘤或结肠癌细胞系；晚期肺癌或卵巢癌患者	环磷酰胺（CTX）	海氏肠球菌从小肠向次级淋巴器官迁移，增加了肿瘤内 CD8/Treg 比值，肠道巴恩斯氏菌在结肠内积聚，促进了肿瘤病灶内产生 IFN-γ 的 γδT 细胞的浸润。两者均受到 Nod2 的抑制。在晚期肺癌和卵巢癌的化疗 – 免疫治疗患者中，海氏肠球菌和肠道巴恩斯氏菌特异性的 Th1 细胞免疫反应对应了更长的无进展生存期	[26]
黑色素瘤小鼠模型	细胞毒 T 淋巴细胞抗原 4（CTLA-4）抗体免疫治疗	多形拟杆菌或脆弱拟杆菌的 T 细胞反应与 CTLA-4 阻断效果相关。实验验证了单用多形拟杆菌、脆弱拟杆菌，和只用脆弱拟杆菌多糖与洋葱伯克霍尔德菌（Burkholderia cepacia）联用，只用脆弱拟杆菌特异性的 T 细胞	[27]
黑色素瘤小鼠模型	程序性死亡配体-1（PD-L1）抗体免疫治疗	双歧杆菌（短双歧杆菌、长双歧杆菌）增强了抗 PD-L1 治疗的疗效[28]	[28]
非小细胞肺癌（NSCLC）患者、非小细胞肾癌（RCC）患者	程序性死亡受体-1（PD-1）抗体免疫治疗	在对 PD-1 治疗有响应的患者中，嗜黏蛋白阿克曼菌的相对丰度较高；通过在无反应的粪便中添加嗜黏蛋白阿克曼菌，在小鼠模型中得到证实	[29]
黑色素瘤患者	PD-1	在对 PD-1 有相应患者中，α 多样性较高，瘤胃球菌 Ruminococcaceae 科的菌的相对丰度较高	[30]

当前的研究聚焦于粪便微生物，很多临床研究正在进行中[31]。

7.1.2 精准治疗的宏基因组学

通过微生物组成成分可以预测对癌症免疫治疗和化疗的反应（表7.2），以及对治疗类风湿性关节炎及2型糖尿病等疾病的药物的响应程度[32,33]。举例来说，与单用甲氨蝶呤相比，使用甲氨蝶呤和雷公藤多苷或只使用雷公藤多苷治疗后，类风湿性关节炎患者唾液中韦荣球菌属的含量大量减少，但这仍然有待更多临床病例的验证[32]。在早期风湿性关节炎患者中，甲氨蝶呤治疗被发现与肺纤维化呈负相关[34]，这一研究没有考虑呼吸道或口腔微生物组。对于恶性胶质瘤，正在进行免疫检查点抑制剂、肽疫苗、树突状细胞疫苗等的临床试验，肠道、口腔和潜在的脑脊液微生物组都可能影响试验结果[35]。肠道微生物色氨酸代谢与代谢性疾病、自身免疫性疾病、神经精神疾病和癌症有关[36-41]，针对色氨酸通路的临床试验应该考虑到微生物组（表7.3、表7.4），这可能会影响药物的剂量、毒性和功效（图7.2）。

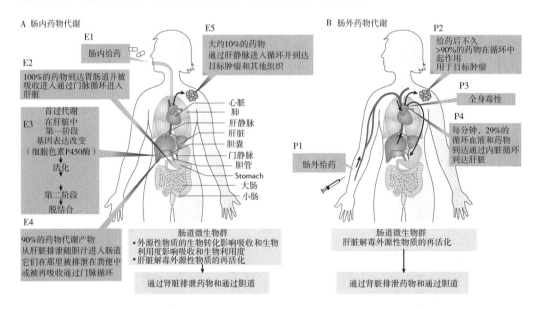

图7.2 药物代谢的主要通路和肠内（如口服）或肠外（如静脉注射）给药后微生物群发挥的作用

A. 肠内药物代谢。口服给药药物（E1）在进入肠道前在胃中停留30~45 min，然后通过门静脉循环（E2）被肝脏吸收。在肠道中，宿主和微生物酶引起药物的代谢变化，能直接结合细菌产物，间接控制肠道吸收。在肝脏中，随着第一和第二阶段的处理（第一阶段代谢；E3），大约90%的口服药物通过胆汁分泌被代谢、破坏或排出（E4）。药物通过胆道分泌进入肠道，可以通过门静脉循环再吸收，也可以从粪便中排出。因此，只有10%的口服药物通过肝静脉进入循环，可以到达靶肿瘤和其他组织（E5）。第一阶段和第二阶段的处理，通过微生物调节参与药物处理的宿主酶的水平，也受到肠道微生物群的影响。B. 肠外药物代谢。静脉给药后（P1）几乎100%的药物进入循环，可到达肿瘤靶细胞（P2），然而，药物通过全身系统性循环，也可能引起毒副反应（P3）。任何未被组织吸收的残留药物都能迅速由肾脏排出体外。每分钟29%的循环药物通过内脏循环（肝动脉、肠系膜动脉和脾动脉）运输到肝脏（P4），此处的处理类似于肠内给药。经过胆汁排泄途径从肝脏分泌到肠道的解毒药物可以被细菌酶重新激活，引起肠道毒性。CYP450：细胞色素P450；GI：胃肠道。来源：Nat Rev Cancer, 2017. https://doi.org/10.1038/nrc.2017.13。

第7章 宏基因组的临床应用

表 7.3 目前研究的 IDO1（吲哚胺 2,3-双加氧酶 1）抑制剂

分子	结构和性质	研究	已发表的研究	正在进行和招募受试者的研究
1-MT-L-Trp（1-methyl-L-tryptophan）	L-色氨酸类似物；非特异性 IDO1 竞争性抑制剂；无论是否有 IDO，增加抗癌药物的疗效，增加体内和体外 KYNA	基础研究	晚期恶性肿瘤；耐受性良好（单一治疗）	I/II 期：乳腺癌（NCT01042535, NCT01792050）、胰腺癌（NCT02077881）、前列腺癌（NCT01560923）、非小细胞肺癌（NCT02460367）、实体瘤（NCT00567931, NCT01191216）、脑肿瘤（NCT04049669, NCT02052648, NCT02502708）、白血病（NCT02835729）和黑色素瘤（NCT03301636, NCT02073123）
1-MT-D-Trp（1-甲基-d-色氨酸，吲哚西莫德）	体外活性低，但体内有效，更倾向于抑制 IDO2；可能通过非靶向作用促进肿瘤生长	癌症（单独或联合使用）		
Epacadostat INCB024360	前药：NLG802 选择性和不可逆的合成 IFN-γ，增加 T 细胞合成），但作为单一治疗缺乏活性；通过体内微生物群和 UGT1A9（AhR 靶）进行代谢	癌症（单独或联合使用）	卵巢癌：无益[48] 肿瘤：耐受性良好，具有抗肿瘤活性[49] 转移性黑色素瘤：无益	I/II 期：胸腺癌（NCT02364076）、鼻咽癌（NCT03196232）、胃癌（NCT03291054）、胰腺癌（NCT03006302）、泌尿膀胱癌（NCT03832673）、非小细胞肺癌（NCT03322566, NCT03322540）、直肠癌（NCT03516708）、黑色素实体瘤（NCT01961115）、肉瘤（NCT03414229）、转移性实体肿瘤（NCT03347123）和肾癌（nct033361865, NCT03374488, NCT03260894, NCT03358472）
5Linrodostat BMS-986205iMS-986	强效、选择性和不可逆的 IDO1 抑制剂，恢复 T 细胞增殖和降低瘤内 L-kyn 高达 90%	癌症	肿瘤：耐受良好（±nivolumab），需要进一步研究疗效	I/II 期：药代动力学（NCT03378310, NCT03312426）和安全性（NCT03192943）、体内试验（NCT04106414）、肝脏（NCT03695250）、胃癌（NCT02935634）、头颈部（NCT03854032）和膀胱癌（NCT03519256）、实体肿瘤（NCT03792750）和膀胱癌（NCT03459222, NCT02658890）III 期：膀胱癌（NCT03661320）、胶质瘤（NCT03661320）、黑色素瘤（NCT03329846）
EOS200271	口服非完整性抑制剂，可以进入大脑	和 PD-1 抑制剂有关的胶质瘤[54,55]	恶性胶质瘤：耐受性良好	

续表

分子	结构和性质	研究	已发表的研究	正在进行和招募受试者的研究
那伏莫德 navoximod, GDC-0919, or NLG-919	中度选择性非竞争性可逆抑制剂；效应T细胞的剂量依赖性活动和增殖；大肿瘤的消退；与吲哚西莫特的协同作用；增加目前药物制剂优化的生存期（±）	癌症[56]	复发性进展性实体肿瘤；耐受性良好并减少胞浆I-基因	I/II期：实体肿瘤（NCT02471846, NCT02048709）
芳香烃受体激动剂拉奎莫德（Laquinimod）	喹啉3-甲酰胺结构类似KYNA；AhR依赖性脑脊髓炎效应；混合结果（II期和III期临床试验——多发性硬化）	亨廷顿综合征 哮喘（已上市） 类风湿性关节炎，多发性硬化症，高尿酸血症		
曲尼司特	芦荟髓鞘再生 抗核抗体的合成类似物	特应性皮炎性银屑病		
苯维莫德（benvitimod）	细胞性二苯乙烯；自由基清除剂；皮肤应用			

表7.4 目前被研究的AhR（芳香烃受体）激动剂和拮抗剂

分子	结构和性质	研究	已发表的研究	正在进行和招募受试者的研究
芳香烃受体激动剂拉奎莫德（Laquinimod）	喹啉胺结构类似KYNA；AhR依赖性脑脊髓炎效应；混合结果（II期和III期临床试验——多发性硬化）	亨廷顿综合征 氏多重克罗恩病 哮喘（已上市）	多发性硬化症：耐受良好，对脑萎缩有显著抑制作用[60,61]（NCT01975298）；克罗恩病：耐受良好，有希望的效果[62]	I/II期：相关多发性硬化症（NCT01047319），亨廷顿（NCT02215616），狼疮性关节炎（NCT01085084），狼疮性肾炎（NCT01085097），克罗恩病（NCT00737932），多发性硬化症复发（NCT01975298）的疗效和安全性
曲尼司特	芦荟髓鞘再生 抗核抗体的合成类似物	类风湿性关节炎，多发性硬化症，高尿酸血症	前列腺癌：对预后有益[63]	I/II期：瓣蛋白（NCT03490708），硬肿症（NCT03512873），结节病（NCT03528070），冻融相关周期综合征（NCT03923140），翼状胬肉（NCT01003613），高尿酸血症（NCT00995618, n01052987），痛风（NCT01109121），类风湿性关节炎（nct0882024）

续表

分子	结构和性质	研究	已发表的研究	正在进行和招募受试者的研究
苯烯莫德（benvitimod）	细菌性二苯乙烯；自由基清除剂；皮肤应用	特应性皮肤炎性皮炎和异位性皮炎银屑病	银屑病和异位性皮肤炎：耐受性良好	Ⅰ/Ⅱ期：1%（抗张力斑块型银屑病）（NCT04042103）；Ⅲ期：1%（斑块型银屑病）
CH223191	竞争性选择性拮抗剂；非HLA配体无拮抗活性	基础研究，但可能有希望治疗胰腺癌	未进行临床实验	
CB7993113	良好的口服生物利用度；在体外阻断肿瘤细胞迁移并降低ER-/PR-/HER2-乳腺癌细胞的侵袭表型	[68,69]		
StemRegenin-1	体外应用；扩增CD34⁺细胞	中性粒细胞血小板 CD34⁺细胞扩增减少症	中性粒细胞血小板 CD34⁺细胞扩增	恶性血液病（nct01474681 和 NCT01930162）中性粒细胞减少症和低血小板计数（NCT03406962）

从机制上讲，微生物组的药物代谢可以激活、失活或毒化药物（图 7.2，表 7.5），并且可以与人类遗传变异一起发挥作用（表 7.6）。值得注意的是，小鼠实验通常是在雄性小鼠中进行的，以减少由雌性小鼠的发情周期而引起的变异性，所以动物实验结果和人类实验结果之间的差异，除了肠型的差异外（第 2 章），还有一部分可能是由于不同性别的微生物和免疫反应的差异引起。

在治疗前进行微生物检测可以节省宝贵时间，并有助于指导用药和其他疗法（图 7.2 ~ 图 7.4）。每个患者的最适剂量取决于其人类基因组和微生物组，微生物组可以激活、灭活或转化多种分子，包括药物（图 7.2，图 7.3；表 7.5，表 7.6）[91,92]。一些与疾病相关的微生物仍然不受现有临床一线药物的影响[32,93]，这意味着需要组合治疗，或新药的深入研发。微生物组的信息也可以用来预先阻断或控制副作用，例如，在癌症免疫治疗期间[94]。如果患者倾向于自发停止使用药物，定期检测微生物组展现出对健康状态的趋近也可能有助于他们继续服药。治疗后的微生物组检测也可以指导医生决定继续或停止用药，亦或转为其他治疗方案。

高强度间歇训练（HIIT）已经在糖尿病早期患者和阿尔茨海默病患者身上试用过[95-98]。考虑到粪便、口腔、阴道微生物群和体质测评结果（例如肌力测试、肺活量）之间的关系等[99-101]，微生物组检测也可以帮助指导更个性化的身体锻炼。此外，运动可能也代表着出汗，出汗除了分泌氯化钠、尿素、乳酸盐、肌酐等之外，还能排出非必需或有毒的微量金属[102]，可能降低疾病风险。

药物 1	药物 2
在有某些微生物的患者中有效	在有某些微生物的患者中有效
根据微生物代谢进行个性化的剂量调整	根据微生物代谢进行个性化的剂量调整
即使对于见效的患者，疾病相关微生物因素仍未被治疗	中等程度副作用（例如肠道，代谢）会经常导致患者放弃服药
轻微副作用（例如皮肤）	长期的缓解
疾病进展，远期复发	

图 7.3　患者合适用药假想病例

使用微生物组信息，促进更个性化地选择药物类型和剂量，使患者更好地遵守处方、减轻副作用，以及更有效地长期管理疾病。

表 7.5 人肠道微生物菌群对药物的选择性修饰

表型效应	微生物修饰	药物亚型	结果	宿主效应	参考文献
激活和再激活	减少	偶氮还原：柳氮磺胺吡啶（SSZ）、巴柳氮、奥沙拉嗪、奥沙拉秦	前药激活：局部 5-氨基水杨酸释放	抗炎治疗	[71]
	脱烷基化	偶氮还原：百浪多息、新百浪多息	抗生素激活	细菌杀灭	[72]
	去结合	N-脱烷基化：胺碘酮	活性代谢产物生物利用度增加	半衰期增加，可能的药物相互作用	[71]
	其他	去葡糖醛酸化：吗啡、可待因	活性代谢产物的再生成	曲线下面积（AUC）增加，肠肝循环	[71]
		去硫酸化：匹可硫酸钠	溶解度增加	通便作用激活	[72]
失活	还原	硝基还原：苯二氮䓬类：硝西泮、氯硝西泮、溴西泮	转变为无活性代谢产物	药物失活，可能的过量干预	[71,72]
		内酯环化：地高辛	转变为无活性代谢产物	治疗窗狭窄	[71]
	脱烷基化	N-脱甲基化：甲苯丙胺	转变为无活性代谢产物	治疗效果降低	[71]
	脱羟基化	P-脱羟基化：L-多巴	L-多巴吸收减少，由幽门螺杆菌引起	治疗效果降低	[71,72]
	蛋白水解	胰岛素、降钙素	治疗活性蛋白质的分解	治疗效果降低	[72]
	乙酰化	N-乙酰化：5-氨基水杨酸	转变为无活性代谢产物	疗效降低，可能的胰腺毒性	[71]
毒化	还原	硝基还原：氯霉素	对氨基苯基-2-吗啡-葡糖醛酸苷氨基-1,3-丙二醇生成（推测）	骨髓毒性	[72]
	脱烷基化	硝基还原：苯二氮䓬类：硝西泮、氯硝西泮、溴西泮	氨基代谢产物生成，失活	致畸性	[71,72]
		N-脱烷基化：溴夫定、索立夫定	额外的溴乙烯基尿嘧啶生成，药物 AUC 降低，与 5-氟尿嘧啶（5-FU）可能致命的累积毒性相互作用	拟杆菌介导的肝毒性	[73]
	去结合	去葡糖醛酸化：伊立替康、双氯芬酸、酮洛芬、吲哚美辛	细胞毒性药物的再形成	腹泻，肠道不适，胃肠道损伤	[71,72]

5-氨基水杨酸。AUC，曲线下面积，表示药物在一段时间内的血浆浓度，因此较高的 AUC 意味着体内药物更多。

表 7.6 具有潜在人类和细菌变异来源的药物

药物	对应人类药物基因	多态性的影响	微生物相关的代谢	微生物代谢的影响	参考文献
华法林	CYP2C9	药物活性改变	维生素 K 生成	微生物组产生不同浓度的维生素 K。微生物组产生的维生素 K 的变化可能改变华法林的代谢	[74-76]
伊立替康	UGT1A1*28"吉尔伯特综合征"	毒性增加	葡糖醛酸化缺陷，排泄的 SN-38G 代谢产物的去葡糖醛酸化	细胞毒性伊立替康的再形成	[77,78]
可待因	CYP2D6	生物转化为吗啡	变异位基因可能导致吗啡-葡糖醛酸苷代谢产物的去葡糖醛酸化吗啡再形成，更高的吗啡	由于肠肝循环导致吗啡-葡糖醛酸苷代谢产物的 AUC 降低，减少或增加速率	[79,80]
吗啡	SLC22A1, OCT1	吗啡清除率降低	排泄的吗啡-葡糖醛酸苷代谢产物的去葡糖醛酸化	吗啡再形成，更高吗啡由于肠肝循环导致吗啡-葡糖醛酸苷代谢产物的 AUC 增加诱导某些铜绿假单胞菌株的毒力	[79,81,82]
对乙酰氨基酚	UGT1A, SULT1A3	葡糖醛酸化速率增加	磺化和由于无意过量导致的肝衰竭风险降低，磺化减少	磺化代谢产物增加，可能对甲酚磺化竞争性抑制	[81,83,84]
辛伐他汀	SLCO1B1	纯合子的辛伐他汀 AUC 增加 221%	未知	假设由于微生物对初级胆汁酸的变化而增加疗效	[75,85-87]
地高辛	ABCB1	AUC 增加可能增加内酯环	还原毒性	AUC 降低，治疗菌狭窄	[88,89]
溴夫定和二氢嘧啶脱氢酶（DYPD）索立夫定	DYPD	与嘧啶类似物	溴乙烯尿嘧啶	药物-药物相互作用增加额外的溴乙烯尿嘧啶生成肝毒性，溴乙烯尿嘧啶阻止 5-FU 的清除	[73,90]

维生素 K 也被称为甲萘醌，它经常出现在京都基因与基因组百科全书（KEGG，京都基因与基因组生物微生物组）分析中（例如，参考文献 [32]）。AUC，曲线下面积，表示药物在一段时间内的血浆浓度。因此较高的 AUC 意味着体内药物更多。5-FU，5-氟尿嘧啶。

来源：Table 2 of Itching R, Kelly L, Predicting and understanding the human Microbiome's impact on pharmacology. Trends Pharmacol Sci 2019;40:495 – 505. http://doi.org/10.1016/j.tips.2019.04.014.

图 7.4 从实验室到临床应用,再从临床反馈到宏基因组研究中的互动

7.2 临床实践启发深入研究

新的药物、新的疫苗、新的食物或添加剂、新的材料和设备正被应用于人体。人类微生物群落可能再次发挥重要作用。正如在第 4 章中所讨论的,医生处于独特的位置,能够找寻到在人体内发生事件的闭环。口腔和粪便微生物组的非侵入性检测可以在胰腺癌高风险人群中展开[图 7.5,致死率极高的胰腺导管腺癌(PDAC)],而组织样本可能在治疗前即可获得,在长期治疗期间可以进行更多的非侵入性检查。(肿瘤组织)转移也可能携带来自原部位的微生物,这些微生物可通过测序追踪它们的进化过程,并在必要时进行局部治疗。许多药物和食物的代谢物最终会进入尿液,目前还不清楚这些代谢物是如何影响此部位微生物组的。

如果精液、子宫内膜/子宫颈、尿液和粪便样本的宏基因组测试可以进入生殖相关科室,那么关于各种类型的微生物与男女生育能力之间的关系,将更值得多加探索[103-107]。除了得到孩子这个结果,生育对健康的长期影响也应该是一个重要的考虑因素(第 8 章)。

肺部感染可以随着时间演变(图 7.6)。宏基因组组装的基因组可以补充传统的方法学,并改进数据库,以便今后更快地进行分析和采取行动。微生物组不同成员之间的宏基因组联系将有助于准确预测每个患者的预后。小鼠实验表明,肺癌的进展需要肺部微生物,通过 γδT 细胞和中性粒细胞起作用[108]。肺腺癌也可以重设肝脏的生物钟[109]。从流行病学角度来看,膳食纤维、十字花科蔬菜和益生菌与肺癌风险降低有关,而男性摄入大量咖啡与吸烟相互作用,预示着更高的风险[110-112]。

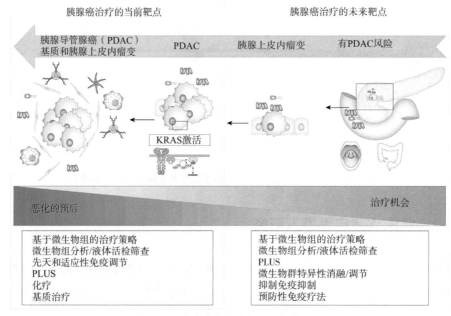

图 7.5 胰腺癌发展阶段的图解说明

目前对胰腺癌的治疗主要集中在早期和晚期的PDAC（浅蓝色子图），这通常预示着预后欠佳。在这个阶段，微生物特异性消融和免疫调节有可能改善胰腺癌的预后，但是治疗效果可能会受到额外的致癌因素的限制，包括肿瘤微环境中KRAS的激活和免疫细胞排斥。事实上，微生物组调节在胰腺癌发育的早期阶段（浅橙色子图）中可能更有影响，在没有不利的肿瘤微环境的情况下，微生物群直接促进肿瘤的发生。微生物组分析、筛查和补充也可能有助于早期的PDAC诊断，并开辟更多的治疗机会。缩写：PDAC，胰腺导管腺癌；PLUS：胰腺消化酶。来源：Trends Cancer, 2019, 5:670-676. https://doi.org/10.1016/j.trecan.2019.10.005.

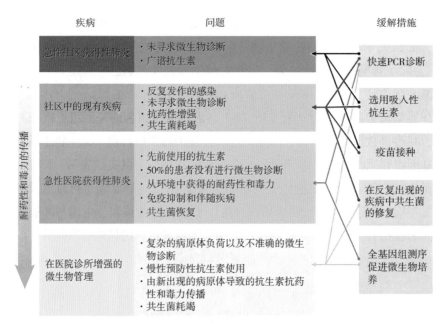

图 7.6 急性和慢性肺部感染的管理

如果PCR是针对正确的病原体设计的,那么它仍然比宏基因组鸟枪法测序更及时。对微生物组和微生物组成员的进化进行详细分析可能会带来更好的长期护理效果。AMR:微生物耐药性;COPD:慢性阻塞性肺病。
来源: Nat Rev Microbiol, 2017. https://doi.org/10.1038/nrmicro.2017.122.

皮肤感染往往也是难以治愈的。不同种的真菌马拉色菌、皮生球菌属缺失和皮肤微生物组中的菌株级进化(第5章),及如何预测特应性皮炎(湿疹)发作值得进一步研究[113]。自身免疫性疾病的不同表现可能与不同的皮肤、黏膜和循环微生物组相关。

针对抗生素耐药性脓肿分枝杆菌的3种混合噬菌体,已用来治疗一位肺移植后15岁的囊性纤维性患者的多发性皮肤损伤。这例患者注射噬菌体治疗后整体见效,且只引起微弱的免疫反应,脓肿分枝杆菌仍可从慢慢愈合皮肤结节中被培养[114]。抗脓肿分枝杆菌噬菌体的抗体也在一个81岁的支气管扩张症患者中被检测到,在这一病例中噬菌体治疗在2个月后不再有效[115]。

对于炎症性肠病中的主要类型——克罗恩病患者,粪便中可能有更多的活泼瘤胃球菌,而溃疡性结肠炎患者粪便中可能有更多的力矩瘤胃球菌[116],并且拟杆菌属减少,通常伴有肠杆菌科的过度生长[117,118]。患者对应疾病的亚型和发病顺序有待确定。有些活泼瘤胃球菌菌株编码的超级抗原能够激活IgA反应[119],这可能会影响肠道微生物组(第2章)。对于一些人来说,在30岁的时候就能看到的更多的力矩瘤胃球菌、B型血和便溏[99],到老年也不会发展成溃疡性结肠炎。2000年初在北美出现的流行病菌株的难辨梭菌在海藻糖的存在下生长迅速[120],但是具有功能正常的肠道微生物的人不需要太担心海藻糖的摄入。治疗炎症性肠病的粪便微生物组移植(FMT)的疗效不如治疗难辨梭菌感染[121-124],相比溃疡性结肠炎,克罗恩病患者的肠道菌在粪便移植时更难被健康供体的粪便菌取代[125]。粪便微生物组可能有助于预测免疫标志物[125,126],口腔微生物组也是免疫紊乱的策源地[127,128],这些都是可能为每个患者选择更有效的治疗方法时应当考虑的。

对于牙医来说,我们是否有一天能够有足够的数据来预测哪些牙齿更容易脱落,而哪些牙齿能保留更久了?口腔正畸治疗也可能改变口腔的通气状况和牙齿表面的唾液保护层。鉴于口腔微生物组与各种疾病之间的联系,医院该如何促进不同科室之间的合作呢?

7.3 利用微生物组知识重新定义现有疾病分类的可能性

疾病的命名往往具有历史的沿袭,就像微生物的命名一样。然而,利用微生物组提供的关键信息层,我们很可能需要对疾病类别进行分组、重组和细分。

没有明显遗传因素的大肠癌(遗传性的如林奇综合征)被称为散发性大肠癌。但是我们现在知道,除了饮食和肥胖的风险因素之外,罪魁祸首可能是一些细菌,而且患者并不需要每一种菌感染。如果有更多关于预后和最佳治疗的证据，根据单独或合并的不同粪便或黏膜细菌，它们很可能被命名为大肠癌和腺瘤的亚型[129-136]。据推测，突变和免疫亚型[137,138]是基因、微生物和环境之间长期相互作用的结果。梭杆菌属，尤其是具核梭杆菌，是近年来研究最多的大肠癌相关菌[139,140]。除了作为腺瘤和癌的生物标志物外，更高数量的组织具核梭杆菌DNA与以下特征有关：近端结肠的肿瘤定植、较高的胰腺癌阶段（较深的侵袭）、较差的肿瘤分化、微卫星不稳定性高、MLH1超甲基化、CpG岛甲基化（CIMP）高和 *BRAF* 突变[141]。具核梭杆菌还与化疗后复发有关系[142]。

另外，结合基础遗传学的知识，我们可能也需要更新对复杂疾病的命名。对于自身免疫性疾病的命名（图7.7）[143]，例如，类风湿性关节炎中没有一种细菌是100%普遍存在的，就像没有一种自身免疫性抗体是100%普遍存在的一样。哪类患者更容易出现骨质流失加快，并且可能需要更积极、更昂贵的治疗，而不是仅能在服用甲氨蝶呤后默默等待[144]？

思考题 7.2

（1）对于结直肠癌的患者，你会如何寻找微小单胞菌、消化链球菌 *Peptostreptococcus stomatis* 或 *P. anaerobius*[130,145]、不解糖卟啉单胞菌，或者大肠埃希菌之间的临床差异？

（2）从现有的结直肠癌分型开始，你是否看到某些分型在特定的人群中出现得更频繁？在治疗前、中、后，你可以收集哪些样本和其他信息？除了手术切除肿瘤之外，你认为如何才能使治疗更有针对性？

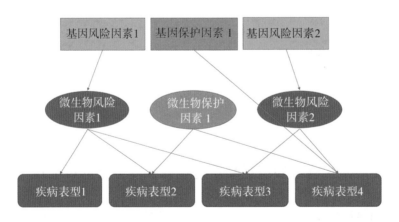

图7.7 可能导致疾病亚型的遗传和微生物组因素的过度简化的示意图

其他临床可用数据，如自身抗体、受影响的淋巴结，也应纳入分类。

7.4 总结

我们已经掌握了微生物组在促进和预防疾病的重要知识，是时候将这些知识运用到临床实践中了。医疗保健专业人士应该考虑是否收集手术样本进行研究，以及是否在接受治疗前后进行宏基因组检测，以评估药物对特定患者的有效性。在设计临床试验时，除了考虑人类基因组的影响，还应该重视患者微生物组的异质性，因为它也会影响药物的效果。此外，结合当前的最佳实践，利用各种疾病的生物标志物，可以进行大规模的人群筛选。我们还需要继续开展临床研究，完善用于诊断和治疗的微生物组模型，以及更深入地理解许多理解各类复杂疾病（图7.4和图7.7）。

原著参考文献

［1］Liao C, Taylor BP, Ceccarani C, Fontana E, Amoretti LA, Wright RJ, et al. Compilation of longitudinal microbiota data and hospitalome from hematopoietic cell transplantation patients. Sci Data 2021;8:71. https://doi.org/10.1038/s41597-021-00860-8.

［2］Khan N, Lindner S, Gomes ALC, Devlin SM, Shah GL, Sung AD, et al. Fecal microbiota diversity disruption and clinical outcomes after auto-HCT: a multicenter observational study. Blood 2021;137:1527–1537. https://doi.org/10.1182/blood.2020006923.

［3］Peled JU, Gomes ALC, Devlin SM, Littmann ER, Taur Y, Sung AD, et al. Microbiota as predictor of mortality in allogeneic hematopoietic-cell transplantation. N Engl J Med 2020;382:822–834. https://doi.org/10.1056/NEJMoa1900623.

［4］Cäcilia Ingham A, Kielsen K, Mordhorst H, Ifversen M, Gottlob Müller K, Johanna Pamp S. Microbiota long-term dynamics and prediction of acute graft-versus-host-disease in pediatric allogeneic stem cell transplantation. MedRxiv 2021. https://doi.org/10.1101/2021.02.19.21252040. 2021.02.19.

［5］Elgarten CW, Tanes C, Lee JJ, Danziger-Isakov LA, Grimley MS, Green M, et al. TITLE: early microbiome and metabolome signatures in pediatric patients undergoing allogeneic hematopoietic cell transplantation. MedRxiv 2021. https://doi.org/10.1101/2021.06.08.21258499. 2021.06.08.21258499.

［6］Moon AM, Singal AG, Tapper EB. Contemporary epidemiology of chronic liver disease and cirrhosis. Clin Gastroenterol Hepatol 2020;18:2650–2666. https://doi.org/10.1016/j.cgh.2019.07.060.

［7］Newsome PN, Cramb R, Davison SM, Dillon JF, Foulerton M, Godfrey EM, et al. Guidelines on the management of abnormal liver blood tests. Gut 2018;67:6–19. https://doi.org/10.1136/gutjnl-2017-314924.

［8］Xie G, Wang X, Zhao A, Yan J, Chen W, Jiang R, et al. Sex-dependent effects on gut microbiota regulate hepatic carcinogenic outcomes. Sci Rep 2017;7:45232. https://doi.org/10.1038/srep45232.

[9] Yuan J, Chen C, Cui J, Lu J, Yan C, Wei X, et al. Fatty liver disease caused by high-alcohol-producing Klebsiella pneumoniae. Cell Metab 2019. https://doi.org/10.1016/j.cmet.2019.08.018.

[10] Gong S, Lan T, Zeng L, Luo H, Yang X, Li N, et al. Gut microbiota mediates diurnal variation of acetaminophen induced acute liver injury in mice. J Hepatol 2018;69:51–59. https://doi.org/10.1016/j.jhep.2018.02.024.

[11] Ramírez-Labrada AG, Isla D, Artal A, Arias M, Rezusta A, Pardo J, et al. The influence of lung microbiota on lung carcinogenesis, immunity, and immunotherapy. Trends Cancer 2020;6:86–97. https://doi.org/10.1016/j.trecan.2019.12.007.

[12] Farrell JJ, Zhang L, Zhou H, Chia D, Elashoff D, Akin D, et al. Variations of oral microbiota are associated with pancreatic diseases including pancreatic cancer. Gut 2012;61:582–588. https://doi.org/10.1136/gutjnl-2011-300784.

[13] Fan X, Alekseyenko AV, Wu J, Peters BA, Jacobs EJ, Gapstur SM, et al. Human oral microbiome and prospective risk for pancreatic cancer: a population-based nested case-control study. Gut 2018;67:120–127. https://doi.org/10.1136/gutjnl-2016-312580.

[14] Zhang X, Hoffman KL, Wei P, Elhor Gbito KY, Joseph R, Li F, et al. Baseline oral microbiome and all-cancer incidence in a cohort of nonsmoking Mexican American women. Cancer Prev Res (Phila) 2021;14:383–392. https://doi.org/10.1158/1940-6207.CAPR-20-0405.

[15] Liu X, Tong X, Zou Y, Lin X, Zhao H, Tian L, et al. Inter-determination of blood metabolite levels and gut microbiome supported by Mendelian randomization. BioRxiv 2020. https://doi.org/10.1101/2020.06.30.181438. 2020.06.30.

[16] Chiu CY, Miller SA. Clinical metagenomics. Nat Rev Genet 2019. https://doi.org/10.1038/s41576-019-0113-7.

[17] Gu W, Deng X, Lee M, Sucu YD, Arevalo S, Stryke D, et al. Rapid pathogen detection by metagenomic next-generation sequencing of infected body fluids. Nat Med 2021;27:115–124. https://doi.org/10.1038/s41591-020-1105-z.

[18] Lennon AM, Buchanan AH, Kinde I, Warren A, Honushefsky A, Cohain AT, et al. Feasibility of blood testing combined with PET-CT to screen for cancer and guide intervention. Science 2020;369. https://doi.org/10.1126/science.abb9601, eabb9601.

[19] Tie J, Cohen JD, Wang Y, Christie M, Simons K, Lee M, et al. Circulating tumor DNA analyses as markers of recurrence risk and benefit of adjuvant therapy for stage III colon cancer. JAMA Oncol 2019. https://doi.org/10.1001/jamaoncol.2019.3616.

[20] Nakamura Y, Taniguchi H, Ikeda M, Bando H, Kato K, Morizane C, et al. Clinical utility of circulating tumor DNA sequencing in advanced gastrointestinal cancer: SCRUM-Japan GI-SCREEN and GOZILA studies. Nat Med 2020. https://doi.org/10.1038/s41591-020-1063-5.

[21] Lo YMD, Han DSC, Jiang P, Chiu RWK. Epigenetics, fragmentomics, and topology of cell-free DNA in liquid biopsies. Science 2021;372. https://doi.org/10.1126/science.aaw3616, eaaw3616.

[22] Witt RG, Blair L, Frascoli M, Rosen MJ, Nguyen QH, Bercovici S, et al. Detection of microbial cell-free DNA in maternal and umbilical cord plasma in patients with chorioamnionitis using next generation sequencing. PLoS One 2020;15. https://doi.org/10.1371/journal.pone.0231239.

[23] Hong DK, Blauwkamp TA, Kertesz M, Bercovici S, Truong C, Banaei N. Liquid biopsy for infectious diseases: sequencing of cell-free plasma to detect pathogen DNA in patients with

invasive fungal disease. Diagn Microbiol Infect Dis 2018;92:210–213. https://doi.org/10.1016/j.diagmicrobio.2018.06.009.

[24] Iida N, Dzutsev A, Stewart CA, Smith L, Bouladoux N, Weingarten RA, et al. Commensal bacteria control cancer response to therapy by modulating the tumor microenvironment. Science 2013;342:967–970. https://doi.org/10.1126/science.1240527.

[25] Viaud S, Saccheri F, Mignot G, Yamazaki T, Daillère R, Hannani D, et al. The intestinal microbiota modulates the anticancer immune effects of cyclophosphamide. Science 2013;342:971–976.

[26] Daillère R, Vétizou M, Waldschmitt N, Yamazaki T, Isnard C, Poirier-Colame V, et al. Enterococcus hirae and Barnesiella intestinihominis facilitate cyclophosphamide-induced therapeutic immunomodulatory effects. Immunity 2016;45:931–943. https://doi.org/10.1016/j.immuni.2016.09.009.

[27] Vétizou M, Pitt JM, Daillère R, Lepage P, Waldschmitt N, Flament C, et al. Anticancer immunotherapy by CTLA-4 blockade relies on the gut microbiota. Science 2015;350:1079–1084. https://doi.org/10.1126/science.aad1329.

[28] Sivan A, Corrales L, Hubert N, Williams JB, Aquino-Michaels K, Earley ZM, et al. Commensal Bifidobacterium promotes antitumor immunity and facilitates anti-PD-L1 efficacy. Science 2015;350:1084–1089. https://doi.org/10.1126/science.aac4255.

[29] Routy B, Le Chatelier E, Derosa L, Duong CPM, Alou MT, Daillère R, et al. Gut microbiome influences efficacy of PD-1–based immunotherapy against epithelial tumors. Science 2017. https://doi.org/10.1126/science.aan3706, eaan3706.

[30] Gopalakrishnan V, Spencer CN, Nezi L, Reuben A, Andrews MC, Karpinets TV, et al. Gut microbiome modulates response to anti–PD-1 immunotherapy in melanoma patients. Science 2017. https://doi.org/10.1126/science.aan4236, eaan4236.

[31] Zhou CB, Zhou YL, Fang JY. Gut microbiota in cancer immune response and immunotherapy. Trends Cancer 2021;7:647–660. https://doi.org/10.1016/j.trecan.2021.01.010.

[32] Zhang X, Zhang D, Jia H, Feng Q, Wang D, Di Liang D, et al. The oral and gut microbiomes are perturbed in rheumatoid arthritis and partly normalized after treatment. Nat Med 2015;21:895–905. https://doi.org/10.1038/nm.3914.

[33] Gu Y, Wang X, Li J, Zhang Y, Zhong H, Liu R, et al. Analyses of gut microbiota and plasma bile acids enable stratification of patients for antidiabetic treatment. Nat Commun 2017;8:1785. https://doi.org/10.1038/s41467-017-01682-2.

[34] Jönsson E, Ljung L, Norrman E, Freyhult E, Ärlestig L, Dahlqvist J, et al. Pulmonary fibrosis in relation to genetic loci in an inception cohort of patients with early rheumatoid arthritis from northern Sweden. Rheumatology (Oxford) 2021. https://doi.org/10.1093/rheumatology/keab441.

[35] Wang H, Xu T, Huang Q, Jin W, Chen J. Immunotherapy for malignant glioma: current status and future directions. Trends Pharmacol Sci 2020;41:123–138. https://doi.org/10.1016/j.tips.2019.12.003.

[36] Li G, Young KD. Indole production by the tryptophanase TnaA in Escherichia coli is determined by the amount of exogenous tryptophan. Microbiology 2013;159:402–410. https://doi.org/10.1099/mic.0.064139-0.

[37] Ye L, Bae M, Cassilly CD, Jabba SV, Thorpe DW, Martin AM, et al. Enteroendocrine cells sense bacterial tryptophan catabolites to activate enteric and vagal neuronal pathways. Cell Host Microbe 2020;29:1–18. https://doi.org/10.1101/2020.06.09.142133.

[38] Qi Q, Li J, Yu B, Moon JY, Chai JC, Merino J, et al. Host and gut microbial tryptophan metabolism and type 2 diabetes: an integrative analysis of host genetics, diet, gut microbiome and circulating metabolites in cohort studies. Gut 2021. https://doi.org/10.1136/gutjnl-2021-324053.

[39] Zhu F, Guo R, Wang W, Ju Y, Wang Q, Ma Q, et al. Transplantation of microbiota from drug-free patients with schizophrenia causes schizophrenia-like abnormal behaviors and dysregulated kynurenine metabolism in mice. Mol Psychiatry 2019. https://doi.org/10.1038/s41380-019-0475-4.

[40] Wlodarska M, Luo C, Kolde R, D'Hennezel E, Annand JW, Heim CE, et al. Indoleacrylic acid produced by commensal peptostreptococcus species suppresses inflammation. Cell Host Microbe 2017;22:25–37.e6. https://doi.org/10.1016/j.chom.2017.06.007.

[41] Dvořák Z, Sokol H, Mani S. Drug mimicry: promiscuous receptors PXR and AhR, and microbial metabolite interactions in the intestine. Trends Pharmacol Sci 2020;41:900–908. https://doi.org/10.1016/j.tips.2020.09.013.

[42] Wirthgen E, Leonard AK, Scharf C, Domanska G. The immunomodulator 1-methyltryptophan drives tryptophan catabolism toward the kynurenic acid branch. Front Immunol 2020;11. https://doi.org/10.3389/fimmu.2020.00313.

[43] Masab M, Saif MW. Telotristat ethyl: proof of principle and the first oral agent in the management of well-differentiated metastatic neuroendocrine tumor and carcinoid syndrome diarrhea. Cancer Chemother Pharmacol 2017;80:1055–1062. https://doi.org/10.1007/s00280-017-3462-y.

[44] Soliman HH, Minton SE, Han HS, Ismail-Khan R, Neuger A, Khambati F, et al. A phase I study of indoximod in patients with advanced malignancies. Oncotarget 2016;7:22928–22938. https://doi.org/10.18632/oncotarget.8216.

[45] Kumar S, Jaipuri FA, Waldo JP, Potturi H, Marcinowicz A, Adams J, et al. Discovery of indoximod prodrugs and characterization of clinical candidate NLG802. Eur J Med Chem 2020;198:112373. https://doi.org/10.1016/j.ejmech.2020.112373.

[46] Boer J, Young-Sciame R, Lee F, Bowman KJ, Yang X, Shi JG, et al. Roles of UGT, P450, and gut microbiota in the metabolism of epacadostat in humans. Drug Metab Dispos 2016;44:1668–1674. https://doi.org/10.1124/dmd.116.070680.

[47] Lewis-Ballester A, Pham KN, Batabyal D, Karkashon S, Bonanno JB, Poulos TL, et al. Structural insights into substrate and inhibitor binding sites in human indoleamine 2,3-dioxygenase 1. Nat Commun 2017;8:1693. https://doi.org/10.1038/s41467-017-01725-8.

[48] Kristeleit R, Davidenko I, Shirinkin V, El-Khouly F, Bondarenko I, Goodheart MJ, et al. A randomised, open-label, phase 2 study of the IDO1 inhibitor epacadostat (INCB024360) versus tamoxifen as therapy for biochemically recurrent (CA-125 relapse)–only epithelial ovarian cancer, primary peritoneal carcinoma, or fallopian tube cancer. Gynecol Oncol 2017;146:484–490. https://doi.org/10.1016/j.ygyno.2017.07.005.

[49] Mitchell TC, Hamid O, Smith DC, Bauer TM, Wasser JS, Olszanski AJ, et al. Epacadostat plus pembrolizumab in patients with advanced solid tumors: phase I results from a multicenter, open-

label phase I/II trial (ECHO-202/KEYNOTE-037). J Clin Oncol 2018;36:3223–3230. https://doi.org/10.1200/JCO.2018.78.9602.

[50] Long GV, Dummer R, Hamid O, Gajewski TF, Caglevic C, Dalle S, et al. Epacadostat plus pembrolizumab versus placebo plus pembrolizumab in patients with unresectable or metastatic melanoma (ECHO-301/KEYNOTE-252): a phase 3, randomised, double-blind study. Lancet Oncol 2019;20:1083–1097. https://doi.org/10.1016/S1470-2045(19)30274-8.

[51] Hunt JT, Balog A, Huang C, Lin T-A, Lin T-A, Maley D, et al. Abstract 4964: Structure, in vitro biology and in vivo pharmacodynamic characterization of a novel clinical IDO1 inhibitor. Exp Mol Ther 2017;4964. https://doi.org/10.1158/1538-7445.AM2017-4964. American Association for Cancer Research.

[52] Luke JJ, Tabernero J, Joshua A, Desai J, Varga AI, Moreno V, et al. BMS-986205, an indoleamine 2, 3-dioxygenase 1 inhibitor (IDO1i), in combination with nivolumab (nivo): updated safety across all tumor cohorts and efficacy in advanced bladder cancer (advBC). J Clin Oncol 2019;37:358. https://doi.org/10.1200/JCO.2019.37.7_suppl.358.

[53] Siu LL, Gelmon K, Chu Q, Pachynski R, Alese O, Basciano P, et al. Abstract CT116: BMS-986205, an optimized indoleamine 2,3-dioxygenase 1 (IDO1) inhibitor, is well tolerated with potent pharmacodynamic (PD) activity, alone and in combination with nivolumab (nivo) in advanced cancers in a phase 1/2a trial. Clin Trials 2017;CT116. https://doi.org/10.1158/1538-7445.AM2017-CT116. American Association for Cancer Research.

[54] Crosignani S, Bingham P, Bottemanne P, Cannelle H, Cauwenberghs S, Cordonnier M, et al. Discovery of a novel and selective indoleamine 2,3-dioxygenase (IDO-1) inhibitor 3-(5-fluoro-1 H -indol-3-yl)pyrrolidine-2,5-dione (EOS200271/PF-06840003) and its characterization as a potential clinical candidate. J Med Chem 2017;60:9617–9629. https://doi.org/10.1021/acs.jmedchem.7b00974.

[55] Reardon DA, Desjardins A, Rixe O, Cloughesy T, Alekar S, Williams JH, et al. A phase 1 study of PF-06840003, an oral indoleamine 2,3-dioxygenase 1 (IDO1) inhibitor in patients with recurrent malignant glioma. Invest New Drugs 2020;38:1784–1795. https://doi.org/10.1007/s10637-020-00950-1.

[56] Sun J, Chen Y, Huang Y, Zhao W, Liu Y, Venkataramanan R, et al. Programmable co-delivery of the immune checkpoint inhibitor NLG919 and chemotherapeutic doxorubicin via a redox-responsive immunostimulatory polymeric prodrug carrier. Acta Pharmacol Sin 2017;38:823–834. https://doi.org/10.1038/aps.2017.44.

[57] Nayak-Kapoor A, Hao Z, Sadek R, Dobbins R, Marshall L, Vahanian NN, et al. Phase Ia study of the indoleamine 2,3-dioxygenase 1 (IDO1) inhibitor navoximod (GDC-0919) in patients with recurrent advanced solid tumors. J Immunother Cancer 2018;6:61. https://doi.org/10.1186/s40425-018-0351-9.

[58] Kaye J, Piryatinsky V, Birnberg T, Hingaly T, Raymond E, Kashi R, et al. Laquinimod arrests experimental autoimmune encephalomyelitis by activating the aryl hydrocarbon receptor. Proc Natl Acad Sci 2016;113:E6145–6152. https://doi.org/10.1073/pnas.1607843113.

[59] Nyamoya S, Steinle J, Chrzanowski U, Kaye J, Schmitz C, Beyer C, et al. Laquinimod supports remyelination in non-supportive environments. Cell 2019;8:1363. https://doi.org/10.3390/

cells8111363.

[60] Ziemssen T, Tumani H, Sehr T, Thomas K, Paul F, Richter N, et al. Safety and in vivo immune assessment of escalating doses of oral laquinimod in patients with RRMS. J Neuroinflammation 2017;14:172. https://doi.org/10.1186/s12974-017-0945-z.

[61] Vollmer TL, Sorensen PS, Selmaj K, Zipp F, Havrdova E, Cohen JA, et al. A randomized placebo-controlled phase III trial of oral laquinimod for multiple sclerosis. J Neurol 2014;261:773–783. https://doi.org/10.1007/s00415-014-7264-4.

[62] D'Haens G, Sandborn WJ, Colombel JF, Rutgeerts P, Brown K, Barkay H, et al. A phase II study of laquinimod in Crohn's disease. Gut 2015;64:1227–1235. https://doi.org/10.1136/gutjnl-2014-307118.

[63] Darakhshan S, Pour AB. Tranilast: a review of its therapeutic applications. Pharmacol Res 2015;91:15–28. https://doi.org/10.1016/j.phrs.2014.10.009.

[64] Smith SH, Jayawickreme C, Rickard DJ, Nicodeme E, Bui T, Simmons C, et al. Tapinarof is a natural AhR agonist that resolves skin inflammation in mice and humans. J Invest Dermatol 2017;137:2110–2119. https://doi.org/10.1016/j.jid.2017.05.004.

[65] Robbins K, Bissonnette R, Maeda-Chubachi T, Ye L, Peppers J, Gallagher K, et al. Phase 2, randomized dose-finding study of tapinarof (GSK2894512 cream) for the treatment of plaque psoriasis. J Am Acad Dermatol 2019;80:714–721. https://doi.org/10.1016/j.jaad.2018.10.037.

[66] Peppers J, Paller AS, Maeda-Chubachi T, Wu S, Robbins K, Gallagher K, et al. A phase 2, randomized dose-finding study of tapinarof (GSK2894512 cream) for the treatment of atopic dermatitis. J Am Acad Dermatol 2019;80:89–98.e3. https://doi.org/10.1016/j.jaad.2018.06.047.

[67] Leja-Szpak A, Góralska M, Link-Lenczowski P, Czech U, Nawrot-Porąbka K, Bonior J, et al. The opposite effect of L-kynurenine and Ahr inhibitor Ch223191 on apoptotic protein expression in pancreatic carcinoma cells (Panc-1). Anticancer Agents Med Chem 2020;19:2079–2090. https://doi.org/10.2174/1871520619666190415165212.

[68] Parks AJ, Pollastri MP, Hahn ME, Stanford EA, Novikov O, Franks DG, et al. In silico identification of an aryl hydrocarbon receptor antagonist with biological activity in vitro and in vivo. Mol Pharmacol 2014;86:593–608. https://doi.org/10.1124/mol.114.093369.

[69] Cheong JE, Sun L. Targeting the IDO1/TDO2–KYN–AhR pathway for cancer immunotherapy—challenges and opportunities. Trends Pharmacol Sci 2018;39:307–325. https://doi.org/10.1016/j.tips.2017.11.007.

[70] Boitano AE, Wang J, Romeo R, Bouchez LC, Parker AE, Sutton SE, et al. Aryl hydrocarbon receptor antagonists promote the expansion of human hematopoietic stem cells. Science 2010;329:1345–1348. https://doi.org/10.1126/science.1191536.

[71] Zhang JJ, Zhang JJ, Wang R. Gut microbiota modulates drug pharmacokinetics. Drug Metab Rev 2018;50:357–368. https://doi.org/10.1080/03602532.2018.1497647.

[72] Wilson ID, Nicholson JK. Gut microbiome interactions with drug metabolism, efficacy, and toxicity. Transl Res 2017;179:204–222. https://doi.org/10.1016/j.trsl.2016.08.002.

[73] Zimmermann M, Zimmermann-Kogadeeva M, Wegmann R, Goodman AL. Separating host and microbiome contributions to drug pharmacokinetics and toxicity. Science 2019;363. https://doi.org/10.1126/science.aat9931, eaat9931.

[74] Violi F, Lip GY, Pignatelli P, Pastori D. Interaction between dietary vitamin K intake and anticoagulation by vitamin K antagonists. Medicine (Baltimore) 2016;95. https://doi.org/10.1097/MD.0000000000002895, e2895.

[75] Ong FS, Deignan JL, Kuo JZ, Bernstein KE, Rotter JI, Grody WW, et al. Clinical utility of pharmacogenetic biomarkers in cardiovascular therapeutics: a challenge for clinical implementation. Pharmacogenomics 2012;13:465–475. https://doi.org/10.2217/pgs.12.2.

[76] Sconce EA, Kamali F. Appraisal of current vitamin K dosing algorithms for the reversal of over-anticoagulation with warfarin: the need for a more tailored dosing regimen. Eur J Haematol 2006;77:457–462. https://doi.org/10.1111/j.0902-4441.2006.t01-1-EJH2957.x.

[77] Guthrie L, Gupta S, Daily J, Kelly L. Human microbiome signatures of differential colorectal cancer drug metabolism. Npj Biofilms Microbiomes 2017;3:27. https://doi.org/10.1038/s41522-017-0034-1.

[78] Lankisch TO, Schulz C, Zwingers T, Erichsen TJ, Manns MP, Heinemann V, et al. Gilbert's syndrome and irinotecan toxicity: combination with UDP-glucuronosyltransferase 1A7 variants increases risk. Cancer Epidemiol Biomarkers Prev 2008;17:695–701. https://doi.org/10.1158/1055-9965.EPI-07-2517.

[79] Sistonen J, Madadi P, Ross CJ, Yazdanpanah M, Lee JW, Landsmeer MLA, et al. Prediction of codeine toxicity in infants and their mothers using a novel combination of maternal genetic markers. Clin Pharmacol Ther 2012;91:692–699. https://doi.org/10.1038/clpt.2011.280.

[80] Kirchheiner J, Schmidt H, Tzvetkov M, Keulen J-T, Lötsch J, Roots I, et al. Pharmacokinetics of codeine and its metabolite morphine in ultra-rapid metabolizers due to CYP2D6 duplication. Pharmacogenomics J 2007;7:257–265. https://doi.org/10.1038/sj.tpj.6500406.

[81] Bairam AF, Rasool MI, Alherz FA, Abunnaja MS, El Daibani AA, Kurogi K, et al. Effects of human SULT1A3/SULT1A4 genetic polymorphisms on the sulfation of acetaminophen and opioid drugs by the cytosolic sulfotransferase SULT1A3. Arch Biochem Biophys 2018;648:44–52. https://doi.org/10.1016/j.abb.2018.04.019.

[82] Tzvetkov MV, dos Santos Pereira JN, Meineke I, Saadatmand AR, Stingl JC, Brockmöller J. Morphine is a substrate of the organic cation transporter OCT1 and polymorphisms in OCT1 gene affect morphine pharmacokinetics after codeine administration. Biochem Pharmacol 2013;86:666–678. https://doi.org/10.1016/j.bcp.2013.06.019.

[83] Clayton TA, Baker D, Lindon JC, Everett JR, Nicholson JK. Pharmacometabonomic identification of a significant host-microbiome metabolic interaction affecting human drug metabolism. Proc Natl Acad Sci 2009;106:14728–14733. https://doi.org/10.1073/pnas.0904489106.

[84] Court MH, Freytsis M, Wang X, Peter I, Guillemette C, Hazarika S, et al. The UDP-glucuronosyltransferase (UGT) 1A polymorphism c.2042C>G (rs8330) is associated with increased human liver acetaminophen glucuronidation, increased UGT1A Exon 5a/5b splice variant mRNA ratio, and decreased risk of unintentional acetaminophen-ind. J Pharmacol Exp Ther 2013;345:297–307. https://doi.org/10.1124/jpet.112.202010.

[85] SLCO1B1. Variants and statin-induced myopathy—a genomewide study. N Engl J Med 2008;359:789–799. https://doi.org/10.1056/NEJMoa0801936.

［86］Voora D, Shah SH, Spasojevic I, Ali S, Reed CR, Salisbury BA, et al. The SLCO1B1*5 genetic variant is associated with statin-induced side effects. J Am Coll Cardiol 2009;54:1609–1616. https://doi.org/10.1016/j.jacc.2009.04.053.

［87］Ramsey LB, Johnson SG, Caudle KE, Haidar CE, Voora D, Wilke RA, et al. The clinical pharmacogenetics implementation consortium guideline for SLCO1B1 and simvastatin-induced myopathy: 2014 update. Clin Pharmacol Ther 2014;96:423–428. https://doi.org/10.1038/clpt.2014.125.

［88］Haiser HJ, Seim KL, Balskus EP, Turnbaugh PJ. Mechanistic insight into digoxin inactivation by Eggerthella lenta augments our understanding of its pharmacokinetics. Gut Microbes 2014;5:233–238. https://doi.org/10.4161/gmic.27915.

［89］Aarnoudse A-JLHJ, Dieleman JP, Visser LE, Arp PP, van der Heiden IP, van Schaik RHN, et al. Common ATP-binding cassette B1 variants are associated with increased digoxin serum concentration. Pharmacogenet Genomics 2008;18:299–305. https://doi.org/10.1097/FPC.0b013e3282f70458.

［90］Diasio RB. Sorivudine and 5-fluorouracil; a clinically significant drug-drug interaction due to inhibition of dihydropyrimidine dehydrogenase. Br J Clin Pharmacol 1998;46:1–4. https://doi.org/10.1046/j.1365-2125.1998.00050.x.

［91］Spanogiannopoulos P, Bess EN, Carmody RN, Turnbaugh PJ. The microbial pharmacists within us: a metagenomic view of xenobiotic metabolism. Nat Rev Microbiol 2016;14:273–287. https://doi.org/10.1038/nrmicro.2016.17.

［92］Hitchings R, Kelly L. Predicting and understanding the human Microbiome's impact on pharmacology. Trends Pharmacol Sci 2019;40:495–505. https://doi.org/10.1016/j.tips.2019.04.014.

［93］Zhu F, Ju Y, Wang W, Wang Q, Guo R, Ma Q, et al. Metagenome-wide association of gut microbiome features for schizophrenia. Nat Commun 2020;11:1612. https://doi.org/10.1038/s41467-020-15457-9.

［94］Chang AE, Golob JL, Schmidt TM, Peltier DC, Lao CD, Tewari M. Targeting the gut microbiome to mitigate immunotherapy-induced colitis in cancer. Trends Cancer 2021;7:583–593. https://doi.org/10.1016/j.trecan.2021.02.005.

［95］Da Silva DE, Grande AJ, Roever L, Tse G, Liu T, Biondi-Zoccai G, et al. High-intensity interval training in patients with type 2 diabetes mellitus: a systematic review. Curr Atheroscler Rep 2019;21:8. https://doi.org/10.1007/s11883-019-0767-9.

［96］Asle Mohammadi Zadeh M, Kargarfard M, Marandi SM, Habibi A. Diets along with interval training regimes improves inflammatory & anti-inflammatory condition in obesity with type 2 diabetes subjects. J Diabetes Metab Disord 2018;17:253–267. https://doi.org/10.1007/s40200-018-0368-0.

［97］Jensen CS, Bahl JM, Østergaard LB, Høgh P, Wermuth L, Heslegrave A, et al. Exercise as a potential modulator of inflammation in patients with Alzheimer's disease measured in cerebrospinal fluid and plasma. Exp Gerontol 2019;121:91–98. https://doi.org/10.1016/j.exger.2019.04.003.

［98］Fiuza-Luces C, Santos-Lozano A, Joyner M, Carrera-Bastos P, Picazo O, Zugaza JL, et al.

Exercise benefits in cardiovascular disease: beyond attenuation of traditional risk factors. Nat Rev Cardiol 2018;15:731–743. https://doi.org/10.1038/s41569-018-0065-1.

[99] Jie Z, Liang S, Ding Q, Li F, Tang S, Wang D, et al. A transomic cohort as a reference point for promoting a healthy gut microbiome. Med Microecol 2021. https://doi.org/10.1016/j.medmic.2021.100039.

[100] Jie Z, Chen C, Hao L, Li F, Song L, Zhang X, et al. Life history recorded in the vagino-cervical microbiome along with multi-omics. Genomics Proteomics Bioinformatics 2021. https://doi.org/10.1016/j.gpb.2021.01.005.

[101] Wilmanski T, Rappaport N, Earls JC, Magis AT, Manor O, Lovejoy J, et al. Blood metabolome predicts gut microbiome α-diversity in humans. Nat Biotechnol 2019. https://doi.org/10.1038/s41587-019-0233-9.

[102] Cohn JR, Emmett EA. The excretion of trace metals in human sweat. Ann Clin Lab Sci 1978;8:270–275.

[103] Lundy SD, Sangwan N, Parekh NV, Selvam MKP, Gupta S, McCaffrey P, et al. Functional and taxonomic dysbiosis of the gut, urine, and semen microbiomes in male infertility. Eur Urol 2021;79:826–836. https://doi.org/10.1016/j.eururo.2021.01.014.

[104] Chen H, Luo T, Chen T, Wang G. Seminal bacterial composition in patients with obstructive and non-obstructive azoospermia. Exp Ther Med 2018. https://doi.org/10.3892/etm.2018.5778.

[105] Chen C, Song X, Wei W, Zhong H, Dai J, Lan Z, et al. The microbiota continuum along the female reproductive tract and its relation to uterine-related diseases. Nat Commun 2017;8:875. https://doi.org/10.1038/s41467-017-00901-0.

[106] Koedooder R, Singer M, Schoenmakers S, Savelkoul PHM, Morré SA, de Jonge JD, et al. The vaginal microbiome as a predictor for outcome of in vitro fertilization with or without intracytoplasmic sperm injection: a prospective study. Hum Reprod 2019;34:1042–1054. https://doi.org/10.1093/humrep/dez065.

[107] Farquhar CM, Bhattacharya S, Repping S, Mastenbroek S, Kamath MS, Marjoribanks J, et al. Female subfertility. Nat Rev Dis Primers 2019;5:1–21. https://doi.org/10.1038/s41572-018-0058-8.

[108] Jin C, Lagoudas GK, Zhao C, Bullman S, Bhutkar A, Hu B, et al. Commensal microbiota promote lung cancer development via γδ T cells. Cell 2019;176:998–1013.e16. https://doi.org/10.1016/j.cell.2018.12.040.

[109] Masri S, Papagiannakopoulos T, Kinouchi K, Liu Y, Cervantes M, Baldi P, et al. Lung adenocarcinoma distally rewires hepatic circadian homeostasis. Cell 2016;165:896–909. https://doi.org/10.1016/j.cell.2016.04.039.

[110] Yang JJ, Yu D, Xiang YB, Blot W, White E, Robien K, et al. Association of dietary fiber and yogurt consumption with lung cancer risk. JAMA Oncol 2019. https://doi.org/10.1001/jamaoncol.2019.4107.

[111] Xie Y, Qin J, Nan G, Huang S, Wang Z, Su Y. Coffee consumption and the risk of lung cancer: an updated meta-analysis of epidemiological studies. Eur J Clin Nutr 2016;70:199–206. https://doi.org/10.1038/ejcn.2015.96.

[112] Vang O. Chemopreventive potential of compounds in cruciferous vegetables. CRC Press; 2005.

[113] Chng KR, Tay ASL, Li C, Ng AHQ, Wang J, Suri BK, et al. Whole metagenome profiling reveals skin microbiome-dependent susceptibility to atopic dermatitis flare. Nat Microbiol 2016;1:16106. https://doi.org/10.1038/nmicrobiol.2016.106.

[114] Dedrick RM, Guerrero-Bustamante CA, Garlena RA, Russell DA, Ford K, Harris K, et al. Engineered bacteriophages for treatment of a patient with a disseminated drug-resistant mycobacterium abscessus. Nat Med 2019;25:730–733. https://doi.org/10.1038/s41591-019-0437-z.

[115] Dedrick RM, Freeman KG, Nguyen JA, Bahadirli-Talbott A, Smith BE, Wu AE, et al. Potent antibody-mediated neutralization limits bacteriophage treatment of a pulmonary Mycobacterium abscessus infection. Nat Med 2021;27(8):1357–1361. https://doi.org/10.1038/s41591-021-01403-9.

[116] Png CW, Lindén SK, Gilshenan KS, Zoetendal EG, McSweeney CS, Sly LI, et al. Mucolytic bacteria with increased prevalence in IBD mucosa augment in vitro utilization of mucin by other bacteria. Am J Gastroenterol 2010;105:2420–2428. https://doi.org/10.1038/ajg.2010.281.

[117] Hebbandi Nanjundappa R, Ronchi F, Wang J, Clemente-Casares X, Yamanouchi J, Sokke Umeshappa C, et al. A gut microbial mimic that hijacks diabetogenic autoreactivity to suppress colitis. Cell 2017;171:655–667.e17. https://doi.org/10.1016/j.cell.2017.09.022.

[118] He Q, Gao Y, Jie Z, Yu X, Laursen JMJM, Xiao L, et al. Two distinct metacommunities characterize the gut microbiota in Crohn's disease patients. Gigascience 2017;6:1–11. https://doi.org/10.1093/gigascience/gix050.

[119] Bunker JJ, Drees C, Watson AR, Plunkett CH, Nagler CR, Schneewind O, et al. B cell superantigens in the human intestinal microbiota. Sci Transl Med 2019;11. https://doi.org/10.1126/scitranslmed.aau9356.

[120] Collins J, Robinson C, Danhof H, Knetsch CW, van Leeuwen HC, Lawley TD, et al. Dietary trehalose enhances virulence of epidemic Clostridium difficile. Nature 2018;553(7688):291–294. https://doi.org/10.1038/nature25178.

[121] Moayyedi P, Surette MG, Kim PT, Libertucci J, Wolfe M, Onischi C, et al. Fecal microbiota transplantation induces remission in patients with active ulcerative colitis in a randomized controlled trial. Gastroenterology 2015;149:102–109.e6. https://doi.org/10.1053/j.gastro.2015.04.001.

[122] Colman RJ, Rubin DT. Fecal microbiota transplantation as therapy for inflammatory bowel disease: a systematic review and meta-analysis. J Crohns Colitis 2014. https://doi.org/10.1016/j.crohns.2014.08.006.

[123] Kelly CP. Fecal microbiota transplantation—an old therapy comes of age. N Engl J Med 2013;368:474–475. https://doi.org/10.1056/NEJMe1214816.

[124] Rossen NG. Fecal microbiota transplantation as novel therapy in gastroenterology: a systematic review. World J Gastroenterol 2015;21:5359. https://doi.org/10.3748/wjg.v21.i17.5359.

[125] Zou M, Jie Z, Cui B, Wang H, Feng Q, Zou Y, et al. Fecal microbiota transplantation results in bacterial strain displacement in patients with inflammatory bowel diseases. FEBS Open Bio 2019. https://doi.org/10.1002/2211-5463.12744.

[126] Schirmer M, Smeekens SP, Vlamakis H, Jaeger M, Oosting M, Franzosa EA, et al. Linking the human gut microbiome to inflammatory cytokine production capacity. Cell 2016;167:1125–

1136.e8. https://doi.org/10.1016/j.cell.2016.10.020.

[127] Atarashi K, Suda W, Luo C, Kawaguchi T, Motoo I, Narushima S, et al. Ectopic colonization of oral bacteria in the intestine drives T H 1 cell induction and inflammation. Science 2017;358:359–365. https://doi.org/10.1126/science.aan4526.

[128] Williams DW, Greenwell-Wild T, Brenchley L, Dutzan N, Overmiller A, Sawaya AP, et al. Human oral mucosa cell atlas reveals a stromal-neutrophil axis regulating tissue immunity. Cell 2021. https://doi.org/10.1016/j.cell.2021.05.013.

[129] Feng Q, Liang S, Jia H, Stadlmayr A, Tang L, Lan Z, et al. Gut microbiome development along the colorectal adenoma–carcinoma sequence. Nat Commun 2015;6:6528. https://doi.org/10.1038/ncomms7528.

[130] Yu J, Feng Q, Wong SHSH, Zhang D, Liang QY, Qin Y, et al. Metagenomic analysis of faecal microbiome as a tool towards targeted non-invasive biomarkers for colorectal cancer. Gut 2017;66:70–78. https://doi.org/10.1136/gutjnl-2015-309800.

[131] Zeller G, Tap J, Voigt AY, Sunagawa S, Kultima JR, Costea PI, et al. Potential of fecal microbiota for early-stage detection of colorectal cancer. Mol Syst Biol 2014;10:766. https://doi.org/10.15252/msb.20145645.

[132] Thomas AM, Manghi P, Asnicar F, Pasolli E, Armanini F, Zolfo M, et al. Metagenomic analysis of colorectal cancer datasets identifies cross-cohort microbial diagnostic signatures and a link with choline degradation. Nat Med 2019;25:667–678. https://doi.org/10.1038/s41591-019-0405-7.

[133] Kasai C, Sugimoto K, Moritani I, Tanaka J, Oya Y, Inoue H, et al. Comparison of human gut microbiota in control subjects and patients with colorectal carcinoma in adenoma: terminal restriction fragment length polymorphism and next-generation sequencing analyses. Oncol Rep 2016;35:325–333. https://doi.org/10.3892/or.2015.4398.

[134] Yachida S, Mizutani S, Shiroma H, Shiba S, Nakajima T, Sakamoto T, et al. Metagenomic and metabolomic analyses reveal distinct stage-specific phenotypes of the gut microbiota in colorectal cancer. Nat Med 2019;25:968–976. https://doi.org/10.1038/s41591-019-0458-7.

[135] Wirbel J, Pyl PT, Kartal E, Zych K, Kashani A, Milanese A, et al. Meta-analysis of fecal metagenomes reveals global microbial signatures that are specific for colorectal cancer. Nat Med 2019;25:679–689. https://doi.org/10.1038/s41591-019-0406-6.

[136] Osman MA, Neoh HM, Ab Mutalib NS, Chin SF, Mazlan L, Raja Ali RA, et al. Parvimonas micra, Peptostreptococcus stomatis, Fusobacterium nucleatum and Akkermansia muciniphila as a four-bacteria biomarker panel of colorectal cancer. Sci Rep 2021;11:2925. https://doi.org/10.1038/s41598-021-82465-0.

[137] Soldevilla B, Carretero-Puche C, Gomez-Lopez G, Al-Shahrour F, Riesco MC, Gil-Calderon B, et al. The correlation between immune subtypes and consensus molecular subtypes in colorectal cancer identifies novel tumour microenvironment profiles, with prognostic and therapeutic implications. Eur J Cancer 2019;123:118–129. https://doi.org/10.1016/j.ejca.2019.09.008.

[138] Komor MA, Bosch LJ, Bounova G, Bolijn AS, Delis-van Diemen PM, Rausch C, et al. Consensus molecular subtype classification of colorectal adenomas. J Pathol 2018;246:266–276. https://doi.org/10.1002/path.5129.

[139] Alexander JL, Scott AJ, Pouncey AL, Marchesi J, Kinross J, Teare J. Colorectal carcinogenesis:

an archetype of gut microbiota-host interaction. Ecancermedicalscience 2018;12:865. https://doi.org/10.3332/ecancer.2018.865.

[140] Slade DJ. New roles for Fusobacterium nucleatum in cancer: target the bacteria, host or both? Trends Cancer 2021;7:185–187. https://doi.org/10.1016/j.trecan.2020.11.006.

[141] Mima K, Nishihara R, Qian ZR, Cao Y, Sukawa Y, Nowak JA, et al. Fusobacterium nucleatum in colorectal carcinoma tissue and patient prognosis. Gut 2016;65:1973–1980. https://doi.org/10.1136/gutjnl-2015-310101.

[142] Yu T, Guo F, Yu Y, Sun T, Ma D, Han J, et al. Fusobacterium nucleatum promotes chemoresistance to colorectal cancer by modulating autophagy. Cell 2017;170:548–563.e16. https://doi.org/10.1016/j.cell.2017.07.008.

[143] Giacomelli R, Afeltra A, Bartoloni E, Berardicurti O, Bombardieri M, Bortoluzzi A, et al. The growing role of precision medicine for the treatment of autoimmune diseases; results of a systematic review of literature and experts' consensus. Autoimmun Rev 2021;20:102738. https://doi.org/10.1016/j.autrev.2020.102738.

[144] Smolen JS, Landewé RBM, Bijlsma JWJ, Burmester GR, Dougados M, Kerschbaumer A, et al. EULAR recommendations for the management of rheumatoid arthritis with synthetic and biological disease-modifying antirheumatic drugs: 2019 update. Ann Rheum Dis 2020;79:685–699. https://doi.org/10.1136/annrheumdis-2019-216655.

[145] Long X, Wong CC, Tong L, Chu ESH, Ho Szeto C, Go MYY, et al. Peptostreptococcus anaerobius promotes colorectal carcinogenesis and modulates tumour immunity. Nat Microbiol 2019;4:2319–2330. https://doi.org/10.1038/s41564-019-0541-3.

第 8 章

微生物档案描绘生命轨迹

摘 要：人体各个部位的正常菌群在全生命周期中承载着重要的信息，它们可以预测未来的疾病风险，包括过敏、月经紊乱、心血管代谢疾病和神经精神疾病等。微生物组将遗传因素和环境因素联系在一起，为精准医学的发展提供了强大的动力，有可能在临床症状出现之前对许多复杂疾病进行预防。为了实现这一目标，需要工程师、科学家以及来自各行各业的专家，来填补目前微生物组研究中存在的信息空白。随着微生物的健康价值日益显现，以及宏基因组学检测价格走低，生物信息工具的速度和准确性也在不断提高，更多的人将会有动力对其微生物组进行采样，这与人类遗传学相比，将会是一个可调节的风险因素。

关键词：多组学，健康管理，预防护理，多基因风险评分（PRS），污染，商业DNA检测，公众科学，膳食纤维，维生素，益生菌

8.1 在重要阶段进行前瞻性微生物采样

从出生到暮年，身体各个部位的微生物群落都蕴藏着重要的信息，它们能够揭示我们未来可能遭遇的疾病风险（图8.1）。早期微生物的定植导致抗原递呈细胞通过肠道、皮肤等途径，以及其他黏膜部位，将微生物抗原转运到胸腺，从而激活 T 细胞，使其能够保护宿主抵御相关病原体的侵袭，同时维持免疫系统的平衡[1-3]。

8.1.1 从出生开始的微生物记录

一些国家的医疗记录令人印象深刻。如果婴儿早产或足月出生，是否会有粪便、口咽和皮肤的微生物样本？母亲在分娩前是否服用了抗生素、注射了催产素（用于刺激分娩）或其他药物[4]？婴儿的微生物组是否能够持续追踪（表8.1）？在极早早产（不足 28 周）的婴儿中，不同的分娩模式之间的差异不再是一个大问题。通过阴道分娩出生的婴儿更容易被拟杆菌和双歧杆菌定植[19,28,29]。那么父亲和祖父母呢？还有哪些自然环境中的有益菌呢[30]？

图 8.1 对不同年龄阶段的微生物进行监控

表 8.1 出生以来的微生物记录

年龄	事件	口腔	粪便	呼吸道	皮肤	需要关注的症状	参考文献
0	出生或婴儿阶段是否接触抗生素	?	Y		?	儿童肥胖；接种疫苗后抗体诱导受损，但T细胞反应增强	[5-7]
0	在出生时接触催产素	?	Y		?	出生前的注射对内源性催产素构成了补充，可能减少焦虑和自闭症的风险，或者减少过度饮食和肥胖的风险？	[8-10]
1~2月	早产，剖宫产，抗生素，缺乏兄弟姐妹，缺乏狗等		Y	Y	?	在接下来几年哮喘风险增加	[11-14]
任何年龄	疫苗	?	Y			根据微生物组预测疫苗接种效果，如果可能的话使用其他方法	[14]
2~6月	因毛细支气管炎住院治疗，并伴有更多的鼻炎和链球菌感染	Y	?	Y		在3岁前发生经常性喘鸣	[15]
3~6月	牛奶过敏	?	Y			预期出现幼儿期自愈的婴儿过敏	[16-18]
0~12月	缺少母乳喂养	Y	Y	?		缺乏来自母亲的双歧杆菌，免疫发育存在显著差异	[19-22]
4~11月	引入固体食物	Y	Y		?	在停止母乳喂养前，提前接触固体食物，以防止食物过敏	[23,24]
4~12月	更规律的睡眠和进食	?	Y			建立肠道微生物昼夜节律	[19] 第2章示例2、示例3

续表

年龄	事件	口腔	粪便	呼吸道	皮肤	需要关注的症状	参考文献
童年到青少年时期	乡村环境	Y	Y	Y	Y	微生物组成成分的差异可能造成长期后果	[25,27]

这些关联不一定是因果关系（第6章），需要进一步阐明。这并不是一个详尽的列表，但它可以说明需要考虑的条件和样本的范围。Y：强烈推荐；？：需要证据验证。

来自母体的阴道微生物，包括细菌性阴道病（BV）相关的阴道加德菌（现在正式归入双歧杆菌属，称为阴道双歧杆菌[31]）和阴道阿波菌（Fannyhessea vaginae）可以在分娩后的第一周在婴儿肠道中被检出[32]。从羊水中出来进入干燥的环境，环境温度是否影响新生儿呼吸道微生物组、皮肤微生物组和肠道微生物组的成熟？家里是否有宠物[33,34]？在建立稳定的昼夜节律前后，婴儿每日睡眠时间如何（表8.1）[35]？所有这些问题都值得探索。

新生儿的通气道是为了吸吮和呼吸而特别优化的持续通气道，喉头位于嘴巴的高处，与黑猩猩类似。随着婴儿的生长，喉部逐渐向下移动，变为成年人的喉（第3章，图3.5），开始吞咽更多的固体食物而不是母乳（吞咽阻断呼吸），并开始说话[36,37]。我们尚不清楚，这一系列的改变对口腔、呼吸和肠道微生物有何影响。当婴儿开始产生自己的抗体，而不是依赖母亲通过胎盘和母乳（有时是从通过口腔）传递的抗体时，抗体和微生物是如何共同进化成适宜的相互关系（弱结合有助于保存共生群落，第2章），以有效地保护婴儿免受病原体的侵害[38,39]？婴儿肝脏从胚胎期的造血功能转变为与成人类似的（昼夜节律）蛋白合成和免疫功能。除了人类基因组编码的淀粉酶，当我们咀嚼淀粉类食物，如大米时，感受到的甜味是否让我们想到口腔和胃肠道微生物组也会影响味觉和食物偏好？例如，能够分解果胶的多形拟杆菌可能增加我们对柑橘和豌豆的喜爱[42]。

为了确定健康婴儿微生物组的基线，需要考虑不同族群的差异[43,44]。正如在第2章所述，人类遗传学对微生物组的构成有一定的影响。在非洲和东亚人群中，粪便中普雷沃氏菌而不是拟杆菌属为主的等位基因，相比欧洲人群中更为常见[45,46]。在没有哮喘的健康儿童的呼吸道中，普雷沃氏菌也更为普遍[47]。在亚洲、非洲或拉丁美洲的一些地区，双歧杆菌在婴儿肠道微生物组中并不常见[48,49]；除了乳糖酶LCT之外，还有多个与双歧杆菌相关的等位基因[45,46]。变形杆菌如大肠埃希菌和克雷伯氏菌能够编码可代谢磺基喹诺酮（6-脱氧-6-硫葡萄糖）的酶，磺基喹诺酮至少占植物叶片干重的10mg/g[50,51]。当地的谷物、土壤和水可能是影响微生物群落的微量元素和重金属的重要来源[46,52-54]（图8.1）。像美国这样的发达国家的水质不一定都符合健康标准[55]。

8.1.2 对于健康微生物群落有影响的当前和历史事件

对于育龄女性来说，可以采集到许多珍贵的样品（表8.2）。影响阴道菌群的因素从青春期跨越到了绝经期。接种疫苗远不足以预防所有的感染和肿瘤。在第一次怀孕期间收集的微生物组样本也可以在下一次怀孕时用于更好地照顾微生物。但是母亲的微生物组对之后的孩子来说是不同的[54]，阴道里的卷曲乳杆菌少了，肠道（可能还有羊水[69]）里的普雷沃菌多了。孕妇粪便微生物组中的普雷沃菌与孩子较低的食物过敏风险有关[70]。怀孕前较低的BMI与母乳中较多的双歧杆菌有关[71]。妊娠糖尿病的粪便微生物组标志物与T2D的部分相似，与T2D的家族史一致，并且在怀孕期间患有妊娠糖尿病的妇女中，T2D的风险增加[73,74]。痛经的女性，会展示出子宫颈微生物组中假单胞菌、不动杆菌和莫拉氏菌较多，而血浆组氨酸水平较低，已婚妇女不动杆菌和莫拉氏菌的相对丰度确实较低[54]。体外人工授精现在相当普遍。阴道、口腔和肠道微生物组除了有助于提高受孕和足月妊娠的成功率外（表8.2），也可能包含疾病风险的信息。例如，我们能否根据粪便和母乳微生物组来预测乳腺癌的风险（第2章，表2.1；第4章，图4.2）？在每次怀孕后恢复粪便、子宫颈、泌尿道和其他微生物组，对于有些人可能相比其他人更困难，需要付出额外的努力。

表8.2 对于女性的微生物记录

事件	生殖道	口腔	粪便	尿液	母乳	微生物	参考文献
月经周期	Y	Y				月经期间惰性乳杆菌和细菌性阴道病相关细菌增多；分泌期卷曲乳杆菌增多 月经前细菌多，口臭；链球菌？	[54,55]
初次性交	Y					月经后更容易患细菌性阴道病；较晚的初次性行为与短双歧杆菌有关	[54,57]
避孕	Y					未避孕者有更多的阴道加德纳杆菌，口服避孕药似乎与子宫颈惰性乳杆菌、微小脲原体和丛毛单胞菌有关，以及与粪便微生物组有关	[54]
怀孕	Y		?			妊娠期乳酸杆菌增多	[58]
体外受精	Y		?			检测到子宫内膜（或输卵管液）中乳杆菌属和其他细菌与妊娠成功的相关性	[59-61]
自发流产	Y		?			惰性乳杆菌和自发流产呈负相关	[54]
早产	Y	Y	Y	?		感染梭杆菌的小鼠出现早产 缺少乳酸杆菌的多样性阴道菌群出现早产	[62]
生育	Y		Y			曾经生育过的母亲的卷曲乳杆杆菌减少；建议筛查链球菌血管病（GBS）、解脲支原体等以预防新生儿感染 对于剖宫产的母亲来说，可能有更多与肥胖相关的细菌；妊娠期间和妊娠后建议监测糖尿病和高血压等疾病的风险	[54,56] 需要进一步研究

续表

事件	生殖道	口腔	粪便	尿液	母乳	微生物	参考文献
母乳喂养		Y			Y	恢复到以乳酸杆菌为主的微生物群，也可能转为以惰性乳杆菌为主 在未来几年里监测乳腺癌的风险；双歧杆菌和其他细菌与婴儿体内的能否匹配	[54,56] [63-66]
绝经	Y	Y	Y	Y		保持健康的骨密度，代谢健康和尿液功能，没有 HPV 和其他病原体	需要前瞻性队列研究
子宫肌瘤	Y			Y		更多的惰性乳杆菌而不是卷曲乳杆菌	[67-68]
类风湿关节炎	Y	Y	Y			滑液中的真菌与阴道中的吻合吗？滑膜液中的细菌与口腔和肠道中的细菌相吻合？更有效的药物治疗	第 4 章

这些关联不一定是因果关系（第6章），需要进一步阐明。这并不是一个详尽的列表，但它可以说明需要考虑的条件和样本的范围。Y：强烈推荐；？：需要证据验证。

骨质疏松导致的髋部骨折对绝经后妇女来说可能威胁生命，骨密度（BMD）的粪便微生物组标志物应该成为早期干预的目标，可通过益生菌和茶（需要干预证据）等措施展开[75-78]。阴道代谢产物也可能包括来自化妆品和清洁用品中的化合物，这点与皮肤情况类似[80]。

在美国等发达国家，许多食物过敏病例开始于成年期（表8.3），于女性更常见[81]。目前哮喘、过敏性鼻炎等并发症与肠道和呼吸道微生物相关。

表 8.3 美国成年人中和年龄相关的针对特定食物过敏的汇总

特定食物过敏	患病率，%（95% 置信区间）					
	所有年龄	18~29 岁	30~39 岁	40~49 岁	50~59 岁	60 岁及以上
任何食物过敏	10.8（10.4~11.1）	11.3（10.5~12.2）	12.7（11.8~13.7）	10.0（9.2~10.9）	11.9（11.0~12.8）	8.8（8.2~9.4）
花生	1.8（1.7~1.9）	2.5（2.2~2.8）	2.9（2.5~3.3）	1.8（1.5~2.1）	1.4（1.1~1.7）	0.8（0.7~1.0）
树坚果	1.2（1.1~1.3）	1.6（1.3~1.9）	1.7（1.4~2.1）	1.1（0.9~1.4）	1.2（0.9~1.5）	0.6（0.4~0.7）
核桃	0.6（0.6~0.7）	0.8（0.7~1.1）	0.9（0.7~1.3）	0.6（0.5~0.8）	0.7（0.5~0.9）	0.3（0.2~0.4）
杏仁	0.7（0.6~0.8）	0.9（0.7~1.2）	1.0（0.7~1.3）	0.7（0.6~1.0）	0.7（0.5~0.9）	0.3（0.2~0.4）
榛子	0.6（0.5~0.7）	0.7（0.5~0.9）	0.9（0.6~1.2）	0.6（0.4~0.8）	0.6（0.4~0.8）	0.3（0.2~0.4）
美洲山核桃	0.5（0.5~0.5）	0.6（0.5~0.8）	0.8（0.5~1.1）	0.6（0.5~0.8）	0.6（0.4~0.8）	0.5（0.4~0.6）
腰果	0.5（0.5~0.6）	0.8（0.6~1.0）	0.8（0.6~1.1）	0.5（0.4~0.7）	0.5（0.3~0.7）	0.2（0.1~0.3）
开心果	0.4（0.3~0.5）	0.6（0.4~0.8）	0.6（0.4~0.8）	0.5（0.3~0.7）	0.4（0.3~0.6）	0.1（0.1~0.2）
其他树坚果	0.2（0.1~0.2）	0.1（0.1~0.2）	0.1（0.0~0.2）	0.3（0.2~0.6）	0.2（0.1~0.5）	0.1（0.1~0.2）
牛奶	1.9（1.8~2.1）	2.4（2.0~2.9）	2.3（1.9~2.8）	2.0（1.6~2.4）	1.9（1.6~2.2）	1.9（1.6~2.2）
贝类	2.9（2.7~3.1）	2.8（1.5~2.1）	3.6（3.1~4.2）	2.5（2.2~3.0）	3.3（2.8~3.8）	2.6（2.2~3.0）
虾	1.9（1.8~2.1）	1.8（1.5~2.1）	2.5（2.1~3.0）	1.8（1.4~2.1）	2.2（1.8~2.6）	1.6（1.3~1.9）
龙虾	1.3（1.2~1.4）	1.2（1.0~1.5）	1.6（1.3~2.0）	1.3（1.0~1.5）	1.4（1.1~1.7）	1.1（0.9~1.3）
螃蟹	1.3（1.2~1.5）	1.2（1.0~1.5）	1.6（1.3~2.0）	1.3（1.0~1.6）	1.6（1.3~2.0）	1.1（1.3~1.9）

续表

特定食物过敏	患病率，%（95% 置信区间）					
	所有年龄	18~29 岁	30~39 岁	40~49 岁	50~59 岁	60 岁及以上
软体动物	1.6（1.4~1.7）	1.6（1.3~2.0）	2.0（1.7~2.5）	1.3（1.1~1.7）	1.7（1.4~2.0）	1.2（1.0~1.5）
其他贝类	0.3（0.2~0.3）	0.3（0.1~0.5）	0.1（0.9~1.3）	0.3（0.2~0.4）	0.3（0.2~0.5）	0.3（0.2~0.4）
鸡蛋	0.8（0.7~0.9）	1.1（0.7~1.5）	1.1（0.9~1.3）	0.7（0.5~0.9）	0.8（0.6~1.1）	0.5（0.3~0.7）
鳍鱼	0.9（0.8~1.0）	1.1（0.9~1.4）	1.0（0.8~1.2）	0.8（0.6~1.1）	1.0（0.7~1.3）	0.6（0.4~0.7）
小麦	0.8（0.7~0.9）	1.0（0.7~1.3）	1.0（0.8~1.3）	0.8（0.6~1.0）	0.7（0.5~0.9）	0.6（0.4~0.8）
大豆	0.6（0.5~0.7）	0.7（0.5~0.9）	0.8（0.6~1.0）	0.6（0.5~0.8）	0.5（0.4~0.7）	0.4（0.3~0.6）
芝麻	0.2（0.2~0.3）	0.3（0.2~0.4）	0.3（0.2~0.5）	0.2（0.1~0.4）	0.3（0.2~0.5）	0.1（0.0~0.2）

来源：JAMA Netw Open, 2019, 2:e185630. https://doi.org/10.1001/jamanetworkopen.2018.5630.

我们对男性微生物的了解还不多。男婴在肠道微生物在组成和功能能力方面已经与女婴不同[19]。有一个位于 NEGR1 和 LINC01360 之间的基因多态性，它与自闭症和精神分裂症相关，而且只有在男性 M-GWAS（宏基因组全基因组关联研究）中被证明与肠道细菌氨基酸球菌（例如肠道氨基酸球菌）有关[45,82]。高尿酸血症（血液中尿酸水平过高）和痛风在男性中更为常见，肠道和口腔微生物也起了很大的作用[45,83,84]。精液中的厌氧球菌和普雷沃菌与精子质量低有关，而假单胞菌与总精子数有关[85,86]。男性不育症的亚型之间也存在微生物学差异，例如，非梗阻性和梗阻性无精子症（没有精子产生），精索静脉曲张（血管扩张）和无精子症（血管扩张）。葡萄球菌被报道在前列腺癌组织中富集[88]；葡萄球菌和比它丰度更高的丙酸痤疮杆菌与慢性前列腺炎相关；在雌激素受体 α（ESR1）敲除小鼠中，葡萄球菌和丙酸痤疮杆菌在精液中都减少了[89]。一个大胆的推测是，可以通过体能训练来调节激素水平，然后观察全身的微生物种群改善。

我们在第 2 章中提到了生物体的衰老过程。为了更好地理解这个过程，我们建议在不同的生命阶段都记录微生物组的变化（图 8.1），并结合其他组学数据，如激素水平、微量元素、体力活动等，这对一般人群也是有益的[53,90]。握力低于同一年龄组的人，是一个已知的心血管事件的流行病学危险因素，与粪便中大肠埃希菌的相对丰度较高[83]相关，孟德尔随机化（第 6 章）显示，粪便中大肠埃希菌相对丰度较高与 2 型糖尿病、心力衰竭、大肠癌的发病风险有关[46]，并可以预测未来几年的疾病风险。睡眠时间过长（＞9 或 10 h）与寿命缩短有流行病学上的联系[91,92]，这是否与睡眠时唾液分泌不足、缺氧以及口腔和肺部微生物的积累有关还有待观察。一些干预措施，例如，膳食纤维（表 8.4）[42,94,95]、乳制品、维生素、高强度间歇训练、早睡等，带来的健康效益，可以很快通过微生物组成及其功能被测定。服用牛奶与肠道菌群的关联除了会增加某些人患肠易激综合征（IBS）的风险外，还可能与牛奶中的雌激素暴露对粪便微生物菌群的影响有关[90]。微生物组与当地生活习惯也有联系，例如茶叶、豆奶、洗浴用的植物等，这些都可以通过更多志愿者参与的研究来验证。

表 8.4　根据可溶性和发酵特征，对自然界中出现的纤维进行分类

纤维类型	链的长度	来源	对 IBS 的潜在益处	对 IBS 的潜在风险
可溶性高发酵寡糖（包括低聚果糖、高聚果糖）	短链碳水化合物	豆类、坚果和种子小麦、黑麦洋葱、大蒜、朝鲜蓟	通便作用：弱通便作用 通过时间：不加速通过时间 细菌平衡：某些微生物群的选择性生长，如双歧杆菌 SCFA：在回肠末端和近端结肠非常迅速地发酵产生 SCFA 气体产生：高	肠易激综合征患者的快速发酵可能导致肠胃胀气和胃肠道症状 一些已开展的关于肠易激综合征的研究结果不一致 [98]
可溶性高度发酵的"纤维"（如抗性淀粉、果胶、瓜尔豆胶和菊粉）	长链碳水化合物	豆类 黑麦面包、大麦 坚实的香蕉 荞麦谷（喀什）、 小米、燕麦 煮熟后冷却 - 意大利面、土豆和大米	通便：温和的通便作用 通过时间：不加速肠道传输。可以减缓小肠的吸收 细菌平衡：增加整体细菌种类，但对双歧杆菌没有选择性 SCFA：在近端结肠快速发酵产生 SCFA。RS 是生产 SCFA 丁酸的优良基质 气体产生：适中	肠易激综合征患者的快速发酵可能导致肠胃胀气和胃肠道症状 在 IBS 中还没有进行精心设计的研究
中等可溶性纤维（洋车前子壳）和燕麦	长链碳水化合物	卵形车前草和燕麦的种子	通便：良好的通便效果 通过时间：确实加快了通过时间 细菌平衡：增加了整体细菌种类，但几乎没有选择性生长的证据 SCFA：沿结肠长度适度发酵产生 SCFA 气体产生：中等	对 IBS 患者的研究表明，对改善腹泻有一些积极的作用 IBS 的一些患者由于排气的副作用而产生了不一致的结果 [99]
不溶于水的缓慢发酵的"纤维"[如麦麸、木质素（亚麻）、水果和蔬菜]	长链碳水化合物	一些蔬菜和水果 麦麸 全麦谷物 黑麦 糙米、全麦面食、 藜麦 亚麻籽	通便：良好的通便效果 通过时间：确实加快通过时间 细菌平衡：增加了整体细菌种类，但几乎没有选择性生长的证据 SCFA：缓慢发酵，沿着结肠的长度产生 SCFA 气体产生：中 - 高	在 IBS 患者中，小麦麸没有被证明是有效的。一个主要的副作用是产生过量的气体和肿胀 [100]。这可能是由于存在的很多果聚糖也与麦麸 [101] 相关 与麦麸相关的症状可能对许多患者是不可接受的

续表

纤维类型	链的长度	来源	对 IBS 的潜在益处	对 IBS 的潜在风险
不溶不发酵的"纤维"（例如：纤维素、苹婆和甲基纤维素）	长链碳水化合物	高纤维谷物坚果、种子水果和蔬菜的表皮	通便：良好的通便效果 通过时间：会加快传递时间 细菌平衡：没有证据表明存在选择性生长 SCFA：发酵不良 气体产生：低	较少的气体产生 这种纤维类型可能对治疗 IBS 患者的便秘有更好效果。然而，很少有精心设计的研究进行

更多信息，参考综述[94]。微生物组是足够多才多艺，这些只是对纤维的一般性分类，特别是对那些"不发酵"的纤维。近年来，有更多新的证据表明，不同类型的膳食纤维能够诱导细菌的不同生长代谢[42, 93]，但是这些纤维的推荐摄入量有时远远超过了营养学专家的建议。来源：Am J Gastroenterol, 2013, 108:718-727. https://doi.org/10.1038/ajg.2013.63.

随着我们为保护地球上的生态系统而采取的措施，空气中的气体和颗粒物、水和土壤中的金属和有机化合物也将发生变化[96,97]。暴露于与气候变化相关的事件，如野火，可能会对微生物组和免疫系统产生长期影响。我们现在记录的微生物组数据可能具有历史的价值。

8.2　从基因风险到疾病预防

我们在第 7 章中讨论了基于宏基因组学的诊断和治疗方法。对于婴儿来说，满月时的呼吸道微生物组可以预测他们在 6 岁时是否会患上哮喘（表 8.3），并且有一些早期干预措施可以针对早产和剖宫产的婴儿（例如，益生菌[102]、微生物定植[103]、皮质醇和儿茶酚胺暴露量，第 6 章，图 6.4）。在 1 型糖尿病的病程中，也有较大规模研究跟踪了人群的微生物组变化[22,104-106]。

许多疾病的趋势在临床症状出现前几十年就已经形成，而微生物组为我们观察这种早期趋势提供了一个关键维度。例如，在一个平均年龄为 30 岁的队列中，我们可以通过粪便细菌和血浆代谢标志物，发现有些人在大肠癌、高尿酸血症（血液中尿酸水平高，但不一定导致痛风）的风险高于其他人且与肉类摄入有关[53]。牙周病和龋齿的多基因风险评分（PRS）对应的曲线下面积（AUC）已超过 0.8，表明有些人需要更加注意其口腔微生物组的护理[84]。对于预防医学来说，乳腺癌、心血管疾病、阿尔茨海默病等的 PRS 打分都很有前景[107]，而粪便或口腔微生物组可以为此增加一个重要的维度。帕金森病和阿尔茨海默病的病情进展可能可以通过适当控制风险因素来减缓[108,109]。干预措施可以是改变饮食习惯、锻炼身体或者戒烟，亡羊补牢是明智选择。代谢综合征和阿尔茨海默病的血浆生物标志物，例如支链氨基酸和酰基肉碱，可以通过腿部肌肉和肾脏代谢[110-112]。经常进行中等强度的体力活动,如快步走,可以降低患癌症和心血管疾病等主要疾病的风险和死亡率[113,114]。肠道微生物组和炎症标志物在短跑后会发生改变[115]。乳制品被认为是痛风的保护因素[90,116]。因此,如果我们通过粪便和(或)口腔微生物组计算出尿酸水平并不低[45,83,84],并且患帕金森病的风险也不高(帕金森病的粪便微生物组研究难免有用药和病程的影响,但仍然可能包含了真正与疾病发生发展有关的微生物)[117,118],在满足上述两个条件时,那么对于帕金森病遗传风险的人,我们还是可以推荐他们食用乳制品[108](译者注:流行病学调查显示,乳制品是帕金森病的风险因素)。

量化膳食信息时,应考虑氮源(如从亚硝酸盐转移到胺类)、氨基酸、金属、维生素等具体成分,而不是依靠西方专家20世纪以来推行的粗略问卷,这对于每个社会环境来说都是一项长期的努力。苯甲酸钠是一种防腐剂,广泛存在于腌制蔬菜和苏打饮料中,最近正在进行临床试验,以研究苯甲酸钠是否有助于治疗精神分裂症患者的抑郁向症状[119](这种症状类似于抑郁症的社会排斥,而非躁狂的症状)。维生素A缺乏的发病率在全球范围内下降(图7.1)。梭状芽孢杆菌能抑制小鼠肠内维生素A的产生,这是断奶诱导T_{reg}产生、肠道成熟的过程;;成年人中梭状芽孢杆菌与血浆维生素A水平相关[23,53,120,121]。除了肠道微生物组,血浆中维生素A水平也与宫颈-阴道微生物组中的惰性乳杆菌丰度呈负相关(从未生育过的女性中有更多的卷曲乳杆菌),这是除了维生素D、激素水平差异[54,123]或一些遗传因素之外,另一个解释美国专家反复报道的少数族裔阴道乳杆菌较少的可能原因[122](译者注:非洲当地样本中乳杆菌似乎并不少)。

体温和每分钟呼吸次数与育龄女性腹腔中的微生物组有关[67]。科学家们根据对月经周期的长期记录,发现月经周期的变化与月亮周期中的亮度(或引力)差异相匹配,这些差异会影响激素和睡眠[124,126]。生育能力强的女性在满月时经期来潮,在最黑暗的日子里排卵[125]。在宫颈微生物组中,月经不调与阴道乳杆菌(L.vaginalis)丰度正相关,而阴道乳杆菌与卷曲乳杆菌有关[54],因此可能随着年龄增长和激素水平降低而减少。

空气污染是导致心血管疾病[96]、肺癌[112]等疾病的重要危险因素,可以在城市尺度上被记录。对于发展中国家来说,燃烧性物质(而不是来自工厂或汽车[97,128,129])仍然是一个主要的污染源,这意味着人类菌群会接触到不同的化学品和颗粒物。

肺活量被证明与粪便微生物组相关[90]。除了体育锻炼,我们是否知道一个人有多少时间花在说话和唱歌?唾液中的IgA和皮质醇被用作生物标志物,以显示欣赏音乐和参加合唱带来的有益效果,催产素也是一个关键的标志物[132]。

可穿戴设备正在不断地发展,未来除了血液以外的体液也可以被更容易地分析。例如,通过出汗可以对葡萄糖、电解质、乳酸、尿酸、金属进行测量等[133,134]。上静脉血氧饱和度和心率通常可由智能手表提供[135]。这些与疾病相关且易于获得的测量方法可以与微生物组记录相匹配分析。

单细胞测序可以用于检测细胞种群和 T 细胞受体、B 细胞受体序列，但是成本太高，无法替代血常规检查中的细胞计数，也无法替代酶联免疫吸附试验（ELISA）中的细胞因子检测。HLA 型别（MHCⅠ，MHCⅡ）是许多疾病的强遗传因素，可以从全基因组测序中获得，而无需额外的实验步骤。研究较少的类似主要组织相容性复合体（MHC）的蛋白质，如 *CD1* 基因和 MR1（MHCⅠ类相关蛋白 1），对于与人类微生物组相关分子的相互作用也很重要[2,136-141]。在小鼠身上发现，黏膜相关恒定 T 细胞（MAIT）的发育是在生命早期（对于小鼠是在 2 ~ 3 周龄时）的一个时间窗口内发生的。共生细菌可以产生维生素 B_2（核黄素）衍生物 5-(2- 氧代亚丙基氨基)-6-D-核糖基氨基尿嘧啶（5-OP-RU）[2]。当这些细菌或代谢产物作用于皮肤或呼吸道时，MR1 将 5-OP-RU 呈递给 MAIT 细胞的 T 细胞受体，诱导胸腺内 MAIT 的发育[2,3]。这些细胞的数量因人而异。

微生物组信息可能有助于缓解由人类基因组商业测试报告的复杂疾病风险引起的伦理争议[142]。家族史是许多复杂疾病（大肠癌、乳腺癌、高压、自身免疫性疾病等）的强流行病学因素，可能反映了遗传和微生物组的共同相似性。通过增加这些可修改的微生物组信息，我们可以更科学地理解微生物信息和遗传信息如何共同影响疾病风险。关键的问题是，我们是否应该根据这些信息采取一些行动？

使用益生菌的临床试验已经在高风险（亚临床）的个体上进行，例如类风湿性关节炎（图 8.2）。微生物组信息（第 4 章，图 4.4；第 7 章）将对免疫参数作出重要的补充。益生菌也正被测试治疗一些疾病[143-145]，若有机会在病例被诊断前开始临床试验，效果可能更佳。

8.3 总结

从出生到老年，人体各部位的微生物群系都含有重要的信息，可以预测未来的疾病风险（图 8.1）。一生一次的经历，或者长期暴露，都可以在人体微生物组中留下痕迹。微生物组将遗传因素和环境因素联系起来，将极大地促进精准医学发展，并有可能在许多复杂疾病临床症状出现之前进行预防。这需要工程师、科学家以及来自各行各业的人们来填充目前微生物组研究中没有被记录的各种信息。随着微生物的健康价值逐渐凸显，宏基因组测序变得更便宜，生物信息学工具变得更快捷便利，更多的人将有动力对其微生物群落进行采样。更可及的送样检测场所可以出于教育目的、作为体检的一部分或更好地与偶发事件相结合（如居家隔离）。

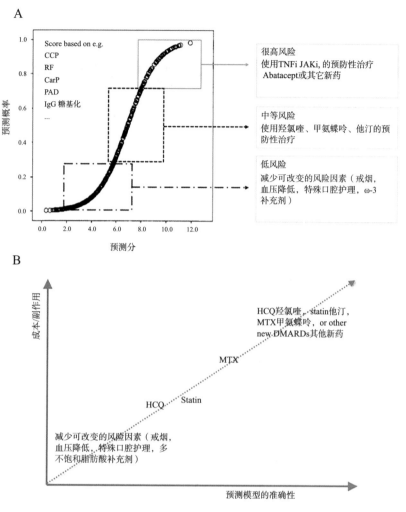

图 8.2 类风湿关节炎的预防

A. 基于风险的预防模式。B. 成本/安全状况与预测模型所要求的准确度之间的线性关系。更昂贵和更不安全的治疗策略对预测模型的准确性要求更高，对近期可能不会患上RA的个体，要避免不必要的检测。DMARD，抗风湿药；HCQ，羟氯喹；甲氨蝶呤；TNFi，肿瘤坏死因子-α抑制剂。来源：Autoimmun Rev, 2020, 19:102506. https://doi.org/10.1016/j.autrev.2020.102506.

思考题 8.1

你所在的国家或地区，你认为最需要的微生物检测是什么？你认为应该在哪里提供检测？你认为人们愿意为这个检测支付多少费用？

思考题 8.2

准备一份简短问卷，供受试者微生物组检测结果之前或之后回答。有些人会比其他人参与意愿更强吗？他们希望微生物组检测能为他们带来什么？你看到现实和预期的差距了吗？你希望如何改进，或者重新聚焦何种领域？

原著参考文献

[1] Zegarra-Ruiz DF, Kim DV, Norwood K, Kim M, Wu W-JH, Saldana-Morales FB, et al. Thymic development of gut-microbiota-specific T cells. Nature 2021;1–5. https://doi.org/10.1038/s41586-021-03531-1.

[2] Legoux F, Bellet D, Daviaud C, El Morr Y, Darbois A, Niort K, et al. Microbial metabolites control the thymic development of mucosal-associated invariant T cells. Science 2019;366. https://doi.org/10.1126/science.aaw2719, eaaw2719.

[3] Constantinides MG, Link VM, Tamoutounour S, Wong AC, Perez-Chaparro PJ, Han SJ, et al. MAIT cells are imprinted by the microbiota in early life and promote tissue repair. Science 2019;366. https://doi.org/10.1126/science.aax6624, eaax6624.

[4] Zhang J, Branch DW, Ramirez MM, Laughon SK, Reddy U, Hoffman M, et al. Oxytocin regimen for labor augmentation, labor progression, and perinatal outcomes. Obstet Gynecol 2011;118:249–256. https://doi.org/10.1097/AOG.0b013e3182220192.

[5] Cho I, Yamanishi S, Cox L, Methé BA, Zavadil J, Li K, et al. Antibiotics in early life alter the murine colonic microbiome and adiposity. Nature 2012;488:621–626. https://doi.org/10.1038/nature11400.

[6] Cox LM, Yamanishi S, Sohn J, Alekseyenko AV, Leung JM, Cho I, et al. Altering the intestinal microbiota during a critical developmental window has lasting metabolic consequences. Cell 2014;158:705–721. https://doi.org/10.1016/j.cell.2014.05.052.

[7] Lynn MA, Tumes DJ, Choo JM, Sribnaia A, Blake SJ, Leong LEX, et al. Early-life antibiotic-driven dysbiosis leads to dysregulated vaccine immune responses in mice. Cell Host Microbe 2018;23:653–660.e5. https://doi.org/10.1016/j.chom.2018.04.009.

[8] Peñagarikano O, Lázaro MT, Lu XH, Gordon A, Dong H, Lam HA, et al. Exogenous and evoked oxytocin restores social behavior in the Cntnap2 mouse model of autism. Sci Transl Med 2015;7:271ra8. https://doi.org/10.1126/scitranslmed.3010257.

[9] Lawson EA. The effects of oxytocin on eating behaviour and metabolism in humans. Nat Rev Endocrinol 2017;13:700–709. https://doi.org/10.1038/nrendo.2017.115.

[10] Ben-Ari Y. Is birth a critical period in the pathogenesis of autism spectrum disorders? Nat Rev Neurosci 2015;16:498–505. https://doi.org/10.1038/nrn3956.

[11] Russell SL, Gold MJ, Hartmann M, Willing BP, Thorson L, Wlodarska M, et al. Early life antibiotic-driven changes in microbiota enhance susceptibility to allergic asthma. EMBO Rep 2012;13:440–447. https://doi.org/10.1038/embor.2012.32.

[12] Pattaroni C, Watzenboeck ML, Schneidegger S, Kieser S, Wong NC, Bernasconi E, et al. Early-life formation of the microbial and immunological environment of the human airways. Cell Host Microbe 2018;24:857–865.e4. https://doi.org/10.1016/j.chom.2018.10.019.

[13] Thorsen J, Rasmussen MA, Waage J, Mortensen M, Brejnrod A, Bønnelykke K, et al. Infant airway microbiota and topical immune perturbations in the origins of childhood asthma. Nat

Commun 2019;10:5001. https://doi.org/10.1038/s41467-019-12989-7.

[14] Lynn DJ, Benson SC, Lynn MA, Pulendran B. Modulation of immune responses to vaccination by the microbiota: implications and potential mechanisms. Nat Rev Immunol 2021. https://doi.org/10.1038/s41577-021-00554-7.

[15] Mansbach JM, Luna PN, Shaw CA, Hasegawa K, Petrosino JF, Piedra PA, et al. Increased Moraxella and Streptococcus species abundance after severe bronchiolitis is associated with recurrent wheezing. J Allergy Clin Immunol 2020;145:518–527.e8. https://doi.org/10.1016/j.jaci.2019.10.034.

[16] Bunyavanich S, Shen N, Grishin A, Wood R, Burks W, Dawson P, et al. Early-life gut microbiome composition and milk allergy resolution. J Allergy Clin Immunol 2016;138:1122–1130. https://doi.org/10.1016/j.jaci.2016.03.041.

[17] Stephen-Victor E, Crestani E, Chatila TA. Dietary and microbial determinants in food allergy. Immunity 2020;53:277–289. https://doi.org/10.1016/j.immuni.2020.07.025.

[18] Rachid R, Stephen-Victor E, Chatila TA. The microbial origins of food allergy. J Allergy Clin Immunol 2021;147:808–813. https://doi.org/10.1016/j.jaci.2020.12.624.

[19] Bäckhed F, Roswall J, Peng Y, Feng Q, Jia H, Kovatcheva-Datchary P, et al. Dynamics and stabilization of the human gut microbiome during the first year of life. Cell Host Microbe 2015;17:690–703. https://doi.org/10.1016/j.chom.2015.04.004.

[20] Verma R, Lee C, Jeun EJ, Yi J, Kim KS, Ghosh A, et al. Cell surface polysaccharides of Bifidobacterium bifidum induce the generation of Foxp3 + regulatory T cells. Sci Immunol 2018;3:eaat6975. https://doi.org/10.1126/sciimmunol.aat6975.

[21] Henrick BM, Rodriguez L, Lakshmikantz T, Pou C, Henckel E, Olin A, et al. Bifidobacteria-mediated immune system imprinting early in life. BioRxiv 2021. https://doi.org/10.1101/2020.10.24.353250.

[22] Vatanen T, Franzosa EA, Schwager R, Tripathi S, Arthur TD, Vehik K, et al. The human gut microbiome in early-onset type 1 diabetes from the TEDDY study. Nature 2018;562:589–594. https://doi.org/10.1038/s41586-018-0620-2.

[23] Al Nabhani Z, Dulauroy S, Marques R, Cousu C, Al Bounny S, Déjardin F, et al. A weaning reaction to microbiota is required for resistance to immunopathologies in the adult. Immunity 2019;50:1276–1288.e5. https://doi.org/10.1016/j.immuni.2019.02.014.

[24] Knoop KA, Gustafsson JK, McDonald KG, Kulkarni DH, Coughlin PE, McCrate S, et al. Microbial antigen encounter during a preweaning interval is critical for tolerance to gut bacteria. Sci Immunol 2017;2. https://doi.org/10.1126/sciimmunol.aao1314, eaao1314.

[25] Lehtimäki J, Karkman A, Laatikainen T, Paalanen L, von Hertzen L, Haahtela T, et al. Patterns in the skin microbiota differ in children and teenagers between rural and urban environments. Sci Rep 2017;7:45651. https://doi.org/10.1038/srep45651.

[26] Ayeni FA, Biagi E, Rampelli S, Fiori J, Soverini M, Audu HJ, et al. Infant and adult gut microbiome and metabolome in rural Bassa and urban settlers from Nigeria. Cell Rep 2018;23:3056–3067. https://doi.org/10.1016/j.celrep.2018.05.018.

[27] Depner M, Taft DH, Kirjavainen PV, Kalanetra KM, Karvonen AM, Peschel S, et al. Maturation of the gut microbiome during the first year of life contributes to the protective farm effect on

childhood asthma. Nat Med 2020. https://doi.org/10.1038/s41591-020-1095-x.

［28］Stokholm J, Thorsen J, Blaser MJ, Rasmussen MA, Hjelmsø M, Shah S, et al. Delivery mode and gut microbial changes correlate with an increased risk of childhood asthma. Sci Transl Med 2020;12. https://doi.org/10.1126/scitranslmed.aax9929, eaax9929.

［29］Selma-Royo M, Calatayud Arroyo M, García-Mantrana I, Parra-Llorca A, Escuriet R, Martínez-Costa C, et al. Perinatal environment shapes microbiota colonization and infant growth: impact on host response and intestinal function. Microbiome 2020;8:167. https://doi.org/10.1186/s40168-020-00940-8.

［30］Kirjavainen PV, Karvonen AM, Adams RI, Täubel M, Roponen M, Tuoresmäki P, et al. Farm-like indoor microbiota in non-farm homes protects children from asthma development. Nat Med 2019;25:1089–1095. https://doi.org/10.1038/s41591-019-0469-4.

［31］Barisic V, Abdelhadi A, Frank A, Riojas MA, Hazbón MH. Reclassification of the bifidobacterium and gardnerella genera. In: ATCC. ASM microbe 2019, San Francisco, California, United States; 2019.

［32］Ferretti P, Pasolli E, Tett A, Asnicar F, Gorfer V, Fedi S, et al. Mother-to-infant microbial transmission from different body sites shapes the developing infant gut microbiome. Cell Host Microbe 2018;24:133–145.e5. https://doi.org/10.1016/j.chom.2018.06.005.

［33］Tun HM, Konya T, Takaro TK, Brook JR, Chari R, Field CJ, et al. Exposure to household furry pets influences the gut microbiota of infant at 3-4 months following various birth scenarios. Microbiome 2017;5:40. https://doi.org/10.1186/s40168-017-0254-x.

［34］Song SJ, Lauber C, Costello EK, Lozupone CA, Humphrey G, Berg-Lyons D, et al. Cohabiting family members share microbiota with one another and with their dogs. Elife 2013;2. https://doi.org/10.7554/eLife.00458, e00458.

［35］Ardura J, Gutierrez R, Andres J, Agapito T. Emergence and evolution of the circadian rhythm of melatonin in children. Horm Res 2003;59:66–72. doi:68571.

［36］Prakash M, Johnny J. Whats special in a child's larynx? J Pharm Bioallied Sci 2015;7:S55–58. https://doi.org/10.4103/0975-7406.155797.

［37］Geddes DT, Chadwick LM, Kent JC, Garbin CP, Hartmann PE. Ultrasound imaging of infant swallowing during breast-feeding. Dysphagia 2010;25:183–191. https://doi.org/10.1007/s00455-009-9241-0.

［38］Pou C, Nkulikiyimfura D, Henckel E, Olin A, Lakshmikanth T, Mikes J, et al. The repertoire of maternal anti-viral antibodies in human newborns. Nat Med 2019. https://doi.org/10.1038/s41591-019-0392-8.

［39］Msallam R, Balla J, Rathore APS, Kared H, Malleret B, Saron WAA, et al. Fetal mast cells mediate postnatal allergic responses dependent on maternal IgE. Science 2020;370:941–950. https://doi.org/10.1126/science.aba0864.

［40］Le Rouzic V, Corona J, Zhou H. Postnatal development of hepatic innate immune response. Inflammation 2011;34:576–584. https://doi.org/10.1007/s10753-010-9265-5.

［41］Nakagaki BN, Mafra K, de Carvalho É, Lopes ME, Carvalho-Gontijo R, de Castro-Oliveira HM, et al. Immune and metabolic shifts during neonatal development reprogram liver identity and function. J Hepatol 2018;69:1294–1307. https://doi.org/10.1016/j.jhep.2018.08.018.

[42] Patnode ML, Beller ZW, Han ND, Cheng J, Peters SL, Terrapon N, et al. Interspecies competition impacts targeted manipulation of human gut bacteria by fiber-derived glycans. Cell 2019;179:59–73.e13. https://doi.org/10.1016/j.cell.2019.08.011.

[43] De Filippo C, Cavalieri D, Di Paola M, Ramazzotti M, Poullet JB, Massart S, et al. Impact of diet in shaping gut microbiota revealed by a comparative study in children from Europe and rural Africa. Proc Natl Acad Sci U S A 2010;107:14691–14696. https://doi.org/10.1073/pnas.1005963107.

[44] Yatsunenko T, Rey FE, Manary MJ, Trehan I, Dominguez-Bello MG, Contreras M, et al. Human gut microbiome viewed across age and geography. Nature 2012;486:222–227. https://doi.org/10.1038/nature11053.

[45] Liu X, Tang S, Zhong H, Tong X, Jie Z, Ding Q, et al. A genome-wide association study for gut metagenome in Chinese adults illuminates complex diseases. Cell Discov 2021;7:9. https://doi.org/10.1038/s41421-020-00239-w.

[46] Liu X, Tong X, Zou Y, Lin X, Zhao H, Tian L, et al. Inter-determination of blood metabolite levels and gut microbiome supported by Mendelian randomization. BioRxiv 2020. https://doi.org/10.1101/2020.06.30.181438. 2020.06.30.

[47] Hilty M, Burke C, Pedro H, Cardenas P, Bush A, Bossley C, et al. Disordered microbial communities in asthmatic airways. PLoS One 2010;5. https://doi.org/10.1371/journal.pone.0008578, e8578.

[48] Schnorr SL, Candela M, Rampelli S, Centanni M, Consolandi C, Basaglia G, et al. Gut microbiome of the Hadza hunter-gatherers. Nat Commun 2014;5:3654. https://doi.org/10.1038/ncomms4654.

[49] Lane AA, McGuire MK, McGuire MA, Williams JE, Lackey KA, Hagen EH, et al. Household composition and the infant fecal microbiome: the INSPIRE study. Am J Phys Anthropol 2019;169:526–539. https://doi.org/10.1002/ajpa.23843.

[50] Denger K, Weiss M, Felux AK, Schneider A, Mayer C, Spiteller D, et al. Sulphoglycolysis in Escherichia coli K-12 closes a gap in the biogeochemical Sulphur cycle. Nature 2014;507:114–117. https://doi.org/10.1038/nature12947.

[51] Roy AB, Hewlins MJE, Ellis AJ, Harwood JL, White GF. Glycolytic breakdown of sulfoquinovose in bacteria: a missing Link in the sulfur cycle. Appl Environ Microbiol 2003;69:6434–6441. https://doi.org/10.1128/AEM.69.11.6434-6441.2003.

[52] Gashu D, Nalivata PC, Amede T, Ander EL, Bailey EH, Botoman L, et al. The nutritional quality of cereals varies geospatially in Ethiopia and Malawi. Nature 2021;594:71–76. https://doi.org/10.1038/s41586-021-03559-3.

[53] Jie Z, Liang S, Ding Q, Li F, Tang S, Wang D, et al. A transomic cohort as a reference point for promoting a healthy gut microbiome. Med Microecol 2021. https://doi.org/10.1016/j.medmic.2021.100039.

[54] Jie Z, Chen C, Hao L, Li F, Song L, Zhang X, et al. Life history recorded in the vagino-cervical microbiome along with multi-omics. Genomics Proteomics Bioinformatics 2021. https://doi.org/10.1016/j.gpb.2021.01.005.

[55] Mueller JT, Gasteyer S. The widespread and unjust drinking water and clean water crisis in the

United States. Nat Commun 2021;12:3544. https://doi.org/10.1038/s41467-021-23898-z.

[56] Dos Santos Santiago GL, Tency I, Verstraelen H, Verhelst R, Trog M, Temmerman M, et al. Longitudinal qPCR study of the dynamics of L. crispatus, L. iners, A. vaginae, (sialidase positive) G. vaginalis, and P. bivia in the vagina. PLoS One 2012;7. https://doi.org/10.1371/journal.pone.0045281, e45281.

[57] Gajer P, Brotman RM, Bai G, Sakamoto J, Schutte UME, Zhong X, et al. Temporal dynamics of the human vaginal microbiota. Sci Transl Med 2012;4:132ra52. https://doi.org/10.1126/scitranslmed.3003605.

[58] Fettweis JM, Serrano MG, Brooks JP, Edwards DJ, Girerd PH, Parikh HI, et al. The vaginal microbiome and preterm birth. Nat Med 2019;25:1012–1021. https://doi.org/10.1038/s41591-019-0450-2.

[59] Koedooder R, Singer M, Schoenmakers S, Savelkoul PHM, Morré SA, de Jonge JD, et al. The vaginal microbiome as a predictor for outcome of in vitro fertilization with or without intracytoplasmic sperm injection: a prospective study. Hum Reprod 2019;34:1042–1054. https://doi.org/10.1093/humrep/dez065.

[60] Schoenmakers S, Laven J. The vaginal microbiome as a tool to predict IVF success. Curr Opin Obstet Gynecol 2020;32:169–178. https://doi.org/10.1097/GCO.0000000000000626.

[61] Pelzer ES, Allan JA, Waterhouse MA, Ross T, Beagley KW, Knox CL. Microorganisms within human follicular fluid: effects on IVF. PLoS One 2013;8. https://doi.org/10.1371/journal.pone.0059062, e59062.

[62] Fardini Y, Chung P, Dumm R, Joshi N, Han YW. Transmission of diverse oral bacteria to murine placenta: evidence for the oral microbiome as a potential source of intrauterine infection. Infect Immun 2010;78:1789–1796. https://doi.org/10.1128/IAI.01395-09.

[63] Urbaniak C, Gloor GB, Brackstone M, Scott L, Tangney M, Reid G. The microbiota of breast tissue and its association with breast cancer. Appl Environ Microbiol 2016;82:5039–5048. https://doi.org/10.1128/AEM.01235-16.

[64] Chambers SA, Townsend SD. Like mother, like microbe: human milk oligosaccharide mediated microbiome symbiosis. Biochem Soc Trans 2020;48:1139–1151. https://doi.org/10.1042/BST20191144.

[65] Pannaraj PS, Li F, Cerini C, Bender JM, Yang S, Rollie A, et al. Association between breast milk bacterial communities and establishment and development of the infant gut microbiome. JAMA Pediatr 2017;171:647–654. https://doi.org/10.1001/jamapediatrics.2017.0378.

[66] Nayfach S, Rodriguez-Mueller B, Garud N, Pollard KS. An integrated metagenomics pipeline for strain profiling reveals novel patterns of bacterial transmission and biogeography. Genome Res 2016;26:1612–1625. https://doi.org/10.1101/gr.201863.115.

[67] Chen C, Song X, Wei W, Zhong H, Dai J, Lan Z, et al. The microbiota continuum along the female reproductive tract and its relation to uterine-related diseases. Nat Commun 2017;8:875. https://doi.org/10.1038/s41467-017-00901-0.

[68] Chen C, Hao L, Wei W, Li F, Song L, Zhang X, et al. The female urinary microbiota in relation to the reproductive tract microbiota. Gigabyte 2020;2020:1–9. https://doi.org/10.46471/gigabyte.9.

[69] Wang J, Zheng J, Shi W, Du N, Xu X, Zhang Y, et al. Dysbiosis of maternal and neonatal

microbiota associated with gestational diabetes mellitus. Gut 2018. https://doi.org/10.1136/gutjnl-2018-315988. gutjnl-2018-315988.

[70] Vuillermin PJ, O'Hely M, Collier F, Allen KJ, Tang MLK, Harrison LC, et al. Maternal carriage of Prevotella during pregnancy associates with protection against food allergy in the offspring. Nat Commun 2020;11:1452. https://doi.org/10.1038/s41467-020-14552-1.

[71] Cortés-Macías E, Selma-Royo M, Martínez-Costa C, Collado MC. Breastfeeding practices influence the breast milk microbiota depending on pre-gestational maternal BMI and weight gain over pregnancy. Nutrients 2021;13. https://doi.org/10.3390/nu13051518.

[72] Crusell MKW, Hansen TH, Nielsen T, Allin KH, Rühlemann MC, Damm P, et al. Gestational diabetes is associated with change in the gut microbiota composition in third trimester of pregnancy and postpartum. Microbiome 2018;6:89. https://doi.org/10.1186/s40168-018-0472-x.

[73] Zhang Y, Xiao CM, Zhang Y, Chen Q, Zhang XQ, Li XF, et al. Factors associated with gestational diabetes mellitus: a Meta-analysis. J Diabetes Res 2021;2021:6692695. https://doi.org/10.1155/2021/6692695.

[74] Hewage SS, Aw S, Chi C, Yoong J. Factors associated with intended postpartum OGTT uptake and willingness to receive preventive behavior support to reduce type 2 diabetes risk among women with gestational diabetes in Singapore: an exploratory study. Nutr Metab Insights 2021;14. https://doi.org/10.1177/11786388211016827. 11786388211016828.

[75] Wang Q, Sun Q, Li X, Wang Z, Zheng H, Ju Y, et al. Linking gut microbiome to bone mineral density: a shotgun metagenomic dataset from 361 elderly women. Gigabyte 2021;2021:1–7. https://doi.org/10.46471/gigabyte.12.

[76] Ohlsson C, Sjögren K. Effects of the gut microbiota on bone mass. Trends Endocrinol Metab 2015;26:69–74. https://doi.org/10.1016/j.tem.2014.11.004.

[77] Yan J, Herzog JW, Tsang K, Brennan CA, Bower MA, Garrett WS, et al. Gut microbiota induce IGF-1 and promote bone formation and growth. Proc Natl Acad Sci U S A 2016;113:E7554–7563. https://doi.org/10.1073/pnas.1607235113.

[78] Zhao H, Chen J, Li X, Sun Q, Qin P, Wang Q. Compositional and functional features of the female premenopausal and postmenopausal gut microbiota. FEBS Lett 2019;593:2655–2664. https://doi.org/10.1002/1873-3468.13527.

[79] Bouslimani A, Porto C, Rath CM, Wang M, Guo Y, Gonzalez A, et al. Molecular cartography of the human skin surface in 3D. Proc Natl Acad Sci U S A 2015;112:E2120–2129. https://doi.org/10.1073/pnas.1424409112.

[80] Kindschuh WF, Baldini F, Liu MC, Gerson KD, Liao J, Lee HH, et al. Preterm birth is associated with xenobiotics and predicted by the vaginal metabolome. BioRxiv 2021. https://doi.org/10.1101/2021.06.14.448190. 2021.06.14.448190.

[81] Gupta RS, Warren CM, Smith BM, Jiang J, Blumenstock JA, Davis MM, et al. Prevalence and severity of food allergies among US adults. JAMA Netw Open 2019;2. https://doi.org/10.1001/jamanetworkopen.2018.5630, e185630.

[82] Zhu F, Ju Y, Wang W, Wang Q, Guo R, Ma Q, et al. Metagenome-wide association of gut microbiome features for schizophrenia. Nat Commun 2020;11:1612. https://doi.org/10.1038/s41467-020-15457-9.

[83] Jie Z, Liang S, Ding Q, Li F, Tang S, Sun X, et al. Disease trends in a young Chinese cohort according to fecal metagenome and plasma metabolites. Med Microecol 2021. https://doi.org/10.1016/j.medmic.2021.100037.

[84] Liu X, Tong X, Zhu J, Tian L, Jie Z, Zou Y, et al. Metagenome-genome-wide association studies reveal human genetic impact on the oral microbiome. bioRxiv 2021. https://doi.org/10.1101/2021.05.06.443017.

[85] Hou D, Zhou X, Zhong X, Settles ML, Herring J, Wang L, et al. Microbiota of the seminal fluid from healthy and infertile men. Fertil Steril 2013;100:1261–1269. https://doi.org/10.1016/j.fertnstert.2013.07.1991.

[86] Lundy SD, Sangwan N, Parekh NV, Selvam MKP, Gupta S, McCaffrey P, et al. Functional and taxonomic dysbiosis of the gut, urine, and semen microbiomes in male infertility. Eur Urol 2021;79:826–836. https://doi.org/10.1016/j.eururo.2021.01.014.

[87] Chen H, Luo T, Chen T, Wang G. Seminal bacterial composition in patients with obstructive and non-obstructive azoospermia. Exp Ther Med 2018. https://doi.org/10.3892/etm.2018.5778.

[88] Cavarretta I, Ferrarese R, Cazzaniga W, Saita D, Lucianò R, Ceresola ER, et al. The microbiome of the prostate tumor microenvironment. Eur Urol 2017;72:625–631. https://doi.org/10.1016/j.eururo.2017.03.029.

[89] Javurek AB, Spollen WG, Ali AMM, Johnson SA, Lubahn DB, Bivens NJ, et al. Discovery of a novel seminal fluid microbiome and influence of estrogen receptor alpha genetic status. Sci Rep 2016;6:23027. https://doi.org/10.1038/srep23027.

[90] Jie Z, Liang S, Ding Q, Li F, Tang S, Wang D, et al. Dairy consumption and physical fitness tests associated with fecal microbiome in a Chinese cohort. Med Microecol 2021.

[91] Svensson T, Saito E, Svensson AK, Melander O, Orho-Melander M, Mimura M, et al. Association of sleep duration with all- and major-cause mortality among adults in Japan, China, Singapore, and Korea. JAMA Netw Open 2021;4(9):e2122837. https://doi.org/10.1001/jamanetworkopen.2021.22837.

[92] Valenzuela PL, Carrera-Bastos P, Gálvez BG, Ruiz-Hurtado G, Ordovas JM, Ruilope LM, et al. Lifestyle interventions for the prevention and treatment of hypertension. Nat Rev Cardiol 2021;18(4):251–275. https://doi.org/10.1038/s41569-020-00437-9.

[93] Kovatcheva-Datchary P, Nilsson A, Akrami R, Lee YS, De Vadder F, Arora T, et al. Dietary fiber-induced improvement in glucose metabolism is associated with increased abundance of prevotella. Cell Metab 2015;22(6):971–982. https://doi.org/10.1016/j.cmet.2015.10.001.

[94] Eswaran S, Muir J, Chey WD. Fiber and functional gastrointestinal disorders. Am J Gastroenterol 2013;108:718–727. https://doi.org/10.1038/ajg.2013.63.

[95] Jie Z, Yu X, Liu Y, Sun L, Chen P, Ding Q, et al. The baseline gut microbiota directs dieting-induced weight loss trajectories. Gastroenterology 2021. https://doi.org/10.1053/j.gastro.2021.01.029.

[96] Al-Kindi SG, Brook RD, Biswal S, Rajagopalan S. Environmental determinants of cardiovascular disease: lessons learned from air pollution. Nat Rev Cardiol 2020;17:656–672. https://doi.org/10.1038/s41569-020-0371-2.

[97] Landrigan PJ, Fuller R, Acosta NJR, Adeyi O, Arnold R, Basu N, et al. The lancet commission on

［98］Bijkerk CJ, Muris JWM, Knottnerus JA, Hoes AW, de Wit NJ. Systematic review: the role of different types of fibre in the treatment of irritable bowel syndrome. Aliment Pharmacol Ther 2004;19:245–251. https://doi.org/10.1111/j.0269-2813.2004.01862.x.

［99］Biesiekierski JR, Rosella O, Rose R, Liels K, Barrett JS, Shepherd SJ, et al. Quantification of fructans, galacto-oligosaccharides and other short-chain carbohydrates in processed grains and cereals. J Hum Nutr Diet 2011;24:154–176. https://doi.org/10.1111/j.1365-277X.2010.01139.x.

［100］Hunt R, Fedorak R, Frohlich J, McLennan C, Pavilanis A. Therapeutic role of dietary fibre. Can Fam Physician 1993;39:897–900 [903–10].

［101］Elia M, Cummings JH. Physiological aspects of energy metabolism and gastrointestinal effects of carbohydrates. Eur J Clin Nutr 2007;61(Suppl 1):S40–74. https://doi.org/10.1038/sj.ejcn.1602938.

［102］Morgan RL, Preidis GA, Kashyap PC, Weizman AV, Sadeghirad B, Chang Y, et al. Probiotics reduce mortality and morbidity in preterm, low birth weight infants: a systematic review and network meta-analysis of randomized trials. Gastroenterology 2020. https://doi.org/10.1053/j.gastro.2020.05.096.

［103］Korpela K, Helve O, Kolho K, Saisto T, Skogberg K, Dikareva E, et al. Maternal fecal microbiota transplantation in cesarean-born infants rapidly restores normal gut microbial development: a proof-of-concept study. Cell 2020;1–11. https://doi.org/10.1016/j.cell.2020.08.047.

［104］Paun A, Yau C, Meshkibaf S, Daigneault MC, Marandi L, Mortin-Toth S, et al. Association of HLA-dependent islet autoimmunity with systemic antibody responses to intestinal commensal bacteria in children. Sci Immunol 2019;4. https://doi.org/10.1126/sciimmunol.aau8125, eaau8125.

［105］Vatanen T, Kostic AD, D'Hennezel E, Siljander H, Franzosa EA, Yassour M, et al. Variation in microbiome LPS immunogenicity contributes to autoimmunity in humans. Cell 2016;165:842–853. https://doi.org/10.1016/j.cell.2016.04.007.

［106］Akil AA-S, Yassin E, Al-Maraghi A, Aliyev E, Al-Malki K, Fakhro KA. Diagnosis and treatment of type 1 diabetes at the dawn of the personalized medicine era. J Transl Med 2021;19:137. https://doi.org/10.1186/s12967-021-02778-6.

［107］Lambert SA, Abraham G, Inouye M. Towards clinical utility of polygenic risk scores. Hum Mol Genet 2019;28:R133–142. https://doi.org/10.1093/hmg/ddz187.

［108］Ascherio A, Schwarzschild MA. The epidemiology of Parkinson's disease: risk factors and prevention. Lancet Neurol 2016;15:1257–1272. https://doi.org/10.1016/S1474-4422(16)30230-7.

［109］Nedergaard M, Goldman SA. Glymphatic failure as a final common pathway to dementia. Science 2020;370:50–56. https://doi.org/10.1126/science.abb8739.

［110］Toledo JB, Arnold M, Kastenmüller G, Chang R, Baillie RA, Han X, et al. Metabolic network failures in Alzheimer's disease: a biochemical road map. Alzheimers Dement 2017;13(9):965–984. https://doi.org/10.1016/j.jalz.2017.01.020.

［111］Overmyer KA, Evans CR, Qi NR, Minogue CE, Carson JJ, Chermside-Scabbo CJ, et al. Maximal oxidative capacity during exercise is associated with skeletal muscle fuel selection and dynamic changes in mitochondrial protein acetylation. Cell Metab 2015;21(3):468–478. https://doi.org/10.1016/j.cmet.2015.02.007.

[112] Jang C, Hui S, Zeng X, Cowan AJ, Wang L, Chen L, et al. Metabolite exchange between mammalian organs quantified in pigs. Cell Metab 2019;30(3):594–606.e3. https://doi.org/10.1016/j.cmet.2019.06.002.

[113] Ruiz-Casado A, Martín-Ruiz A, Pérez LM, Provencio M, Fiuza-Luces C, Lucia A. Exercise and the hallmarks of cancer. Trends Cancer 2017;3:423–441. https://doi.org/10.1016/j.trecan.2017.04.007.

[114] Fiuza-Luces C, Santos-Lozano A, Joyner M, Carrera-Bastos P, Picazo O, Zugaza JL, et al. Exercise benefits in cardiovascular disease: beyond attenuation of traditional risk factors. Nat Rev Cardiol 2018;15:731–743. https://doi.org/10.1038/s41569-018-0065-1.

[115] Motiani KK, Collado MC, Eskelinen JJ, Virtanen KA, LÖyttyniemi E, Salminen S, et al. Exercise training modulates gut microbiota profile and improves endotoxemia. Med Sci Sports Exerc 2020;52:94–104. https://doi.org/10.1249/MSS.0000000000002112.

[116] Kuo CF, Grainge MJ, Zhang W, Doherty M. Global epidemiology of gout: prevalence, incidence and risk factors. Nat Rev Rheumatol 2015;11:649–662. https://doi.org/10.1038/nrrheum.2015.91.

[117] Hopfner F, Künstner A, Müller SH, Künzel S, Zeuner KE, Margraf NG, et al. Gut microbiota in Parkinson disease in a northern German cohort. Brain Res 2017;1667:41–45. https://doi.org/10.1016/j.brainres.2017.04.019.

[118] Bullich C, Keshavarzian A, Garssen J, Kraneveld A, Perez-Pardo P. Gut vibes in Parkinson's disease: the microbiota-gut-brain axis. Mov Disord Clin Pract 2019;6:639–651. https://doi.org/10.1002/mdc3.12840.

[119] Minichino A, Brondino N, Solmi M, Del Giovane C, Fusar-Poli P, Burnet P, et al. The gut-microbiome as a target for the treatment of schizophrenia: a systematic review and meta-analysis of randomised controlled trials of add-on strategies. Schizophr Res 2021;234:1–13. https://doi.org/10.1016/j.schres.2020.02.012.

[120] Grizotte-Lake M, Zhong G, Duncan K, Kirkwood J, Iyer N, Smolenski I, et al. Commensals suppress intestinal epithelial cell retinoic acid synthesis to regulate interleukin-22 activity and prevent microbial dysbiosis. Immunity 2018;49:1103–1115.e6. https://doi.org/10.1016/j.immuni.2018.11.018.

[121] Atarashi K, Tanoue T, Oshima K, Suda W, Nagano Y, Nishikawa H, et al. Treg induction by a rationally selected mixture of Clostridia strains from the human microbiota. Nature 2013;500:232–236. https://doi.org/10.1038/nature12331.

[122] Ravel J, Gajer P, Abdo Z, Schneider GM, Koenig SSK, McCulle SL, et al. Vaginal microbiome of reproductive-age women. Proc Natl Acad Sci U S A 2011;108(Suppl. 1):4680–4687. https://doi.org/10.1073/pnas.1002611107.

[123] Jefferson KK, Parikh HI, Garcia EM, Edwards DJ, Serrano MG, Hewison M, et al. Relationship between vitamin D status and the vaginal microbiome during pregnancy. J Perinatol 2019;39:824–836. https://doi.org/10.1038/s41372-019-0343-8.

[124] Casiraghi L, Spiousas I, Dunster GP, McGlothlen K, Fernández-Duque E, Valeggia C, et al.

Moonstruck sleep: synchronization of human sleep with the moon cycle under field conditions. Sci Adv 2021;7. https://doi.org/10.1126/sciadv.abe0465, eabe0465.

[125] Helfrich-Förster C, Monecke S, Spiousas I, Hovestadt T, Mitesser O, Wehr TA. Women temporarily synchronize their menstrual cycles with the luminance and gravimetric cycles of the moon. Sci Adv 2021;7. https://doi.org/10.1126/sciadv.abe1358, eabe1358.

[126] Taxier LR, Gross KS, Frick KM. Oestradiol as a neuromodulator of learning and memory. Nat Rev Neurosci 2020;21:535–550. https://doi.org/10.1038/s41583-020-0362-7.

[127] Liu Y, Ding H, Ting CS, Lu R, Zhong H, Zhao N, et al. Exposure to air pollution and scarlet fever resurgence in China: a six-year surveillance study. Nat Commun 2020;11:1–13. https://doi.org/10.1038/s41467-020-17987-8.

[128] Daellenbach KR, Uzu G, Jiang J, Cassagnes L, Leni Z, Vlachou A, et al. Sources of particulate-matter air pollution and its oxidative potential in Europe. Nature 2020;587. https://doi.org/10.1038/s41586-020-2902-8.

[129] Alotaibi R, Bechle M, Marshall JD, Ramani T, Zietsman J, Nieuwenhuijsen MJ, et al. Traffic related air pollution and the burden of childhood asthma in the contiguous United States in 2000 and 2010. Environ Int 2019;127:858–867. https://doi.org/10.1016/j.envint.2019.03.041.

[130] Mccraty R, Atkinson M, Rein G, Watkins AD. Music enhances the effect of positive emotional states on salivary IgA. Stress Med 1996;12:167–175. https://doi.org/10.1002/(SICI)1099-1700(199607)12:3<167::AID-SMI697>3.0.CO;2-2.

[131] Kreutz G, Bongard S, Rohrmann S, Hodapp V, Grebe D. Effects of choir singing or listening on secretory immunoglobulin A, cortisol, and emotional state. J Behav Med 2004;27:623–635. https://doi.org/10.1007/s10865-004-0006-9.

[132] Greenberg DM, Decety J, Gordon I. The social neuroscience of music: understanding the social brain through human song. Am Psychol 2021. https://doi.org/10.1037/amp0000819.

[133] Nyein HYY, Bariya M, Kivimäki L, Uusitalo S, Liaw TS, Jansson E, et al. Regional and correlative sweat analysis using high-throughput microfluidic sensing patches toward decoding sweat. Sci Adv 2019;5. https://doi.org/10.1126/sciadv.aaw9906, eaaw9906.

[134] Yang Y, Song Y, Bo X, Min J, Pak OS, Zhu L, et al. A laser-engraved wearable sensor for sensitive detection of uric acid and tyrosine in sweat. Nat Biotechnol 2019. https://doi.org/10.1038/s41587-019-0321-x.

[135] Bayoumy K, Gaber M, Elshafeey A, Mhaimeed O, Dineen EH, Marvel FA, et al. Smart wearable devices in cardiovascular care: where we are and how to move forward. Nat Rev Cardiol 2021;18(8):581–599. https://doi.org/10.1038/s41569-021-00522-7.

[136] Van Rhijn I, Godfrey DI, Rossjohn J, Moody DB. Lipid and small-molecule display by CD1 and MR1. Nat Rev Immunol 2015;15:643–654. https://doi.org/10.1038/nri3889.

[137] Donia MS, Fischbach MA. Small molecules from the human microbiota. Science 2015;349:1254766. https://doi.org/10.1126/science.1254766.

[138] Ma C, Han M, Heinrich B, Fu Q, Zhang Q, Sandhu M, et al. Gut microbiome—mediated bile acid metabolism regulates liver cancer via NKT cells. Science 2018;876. https://doi.org/10.1126/science.aan5931.

[139] Nicolai S, Wegrecki M, Cheng TY, Bourgeois EA, Cotton RN, Mayfield JA, et al. Human T cell

response to CD1a and contact dermatitis allergens in botanical extracts and commercial skin care products. Sci Immunol 2020;5. https://doi.org/10.1126/sciimmunol.aax5430, eaax5430.

［140］An D, Oh SF, Olszak T, Neves JF, Avci FY, Erturk-Hasdemir D, et al. Sphingolipids from a symbiotic microbe regulate homeostasis of host intestinal natural killer T cells. Cell 2014;156:123–133. https://doi.org/10.1016/j.cell.2013.11.042.

［141］Linehan JL, Harrison OJ, Han S-J, Byrd AL, Vujkovic-Cvijin I, Villarino AV, et al. Non-classical immunity controls microbiota impact on skin immunity and tissue repair. Cell 2018;172:784–796.e18. https://doi.org/10.1016/j.cell.2017.12.033.

［142］Becker J, Burik CAP, Goldman G, Wang N, Jayashankar H, Bennett M, et al. Resource profile and user guide of the polygenic index repository. Nat Hum Behav 2021. https://doi.org/10.1038/s41562-021-01119-3.

［143］Pan H, Guo R, Ju Y, Wang Q, Zhu J, Xie Y, et al. A single bacterium resurrects the microbiome-immune balance to protect bones from destruction in a rat model of rheumatoid arthritis. Microbiome 2019;7:107. https://doi.org/10.1186/s40168-019-0719-1.

［144］Pan H, Li R, Li T, Wang J, Liu L. Whether probiotic supplementation benefits rheumatoid arthritis patients: a systematic review and meta-analysis. Engineering 2017;3:115–121. https://doi.org/10.1016/J.ENG.2017.01.006.

［145］O'Toole PW, Marchesi JR, Hill C, Na YC, Kim HS. Next-generation probiotics: the spectrum from probiotics to live biotherapeutics. Nat Microbiol 2017;2:17057. https://doi.org/10.1038/nmicrobiol.2017.57.

致　谢

在过去的一年里，科学家们逐渐习惯了在线会议。节省出更多赶路的时间去思考问题，这是件好事情。

在美国的人类微生物组计划（Human Microbiome Project，HMP）与欧盟的人体肠道宏基因组计划（Metagenome of the Human Intestinal Tract，MetaHIT）尚未启动的许多年前，我才刚刚进入大学的时候，我的微生物学老师，复旦大学的胡宝龙教授就喜欢用"涂布"来形容人们认真洗澡时擦擦洗洗的样子。那时使用的教材，是由胡教授的老师周德庆教授编撰的《微生物学教程》，这本书是中国很多大学生和研究生的微生物学启蒙读物。

钟江教授的病毒学课程让我大开眼界，认识到病毒的神奇之处。杨琰云教授的普通动物学课程生动而全面，课堂上能欣赏到百年前传教士们手绘的精美插图。我的生物化学老师黄伟达教授上课时使用的教材《莱宁格生物化学原理》是我接触到的第一本英文原版教科书。

多年过去，在爱思唯尔出版集团的策划与支持下，使得我有机会出版这样一本关于人类微生物组学这个既古老又年轻的领域的入门教材——*Investigating Human Diseases with The Microbiome: Metagenomics Bench to Bedside*。

许多关于微生物组的宏基因组研究都是基于生物信息学来开展的。我在复旦大学的生物信息学老师钟扬教授热心教育，对西藏大学、上海科技大学等高校的发展均做出贡献。他前些年在内蒙古出差期间因车祸不幸离世。

很幸运我能够在凯斯西储大学攻读博士学位，这所大学坐落于美国的克利夫兰市。那里经济宜居，还经常会举办各种国际文化节庆和艺术活动。

我在博士导师EckhardJankowsky教授的实验室里享受到了非常良好的学习环境。实验室和整个系里的人们都很有趣，也很有协作精神。毕业后我曾在教堂山张毅教授实验室短暂工作过一段时间，随后又去伦敦担任过编辑，这些经历使我进一步认识到，自己并不适合在学术界从事那种专注在某个狭窄领域的传统研究工作。

2012年底一个非常偶然的机会，我访问了华大并与很多在华大工作的年轻人会面、交流。后来便应王俊教授和冯强教授邀请，进入了宏基因组研究这个新兴领域。当时的华大基因公司为了打破Illumina公司的恶意限制，刚刚收购了Complete Genomics公司，急于从源头完成测序技术的升级（当时该项目由徐迅教授主导）。

在各方面资源都相对紧张的情况下，我们还是在宏基因组学领域做出了很好的成绩，始终没有忘记造福人类健康的初心。

杨焕明院士对宏基因组学有着很高的期待。几年前，他曾将宏基因组学技术比作显微镜发明，认为这是微生物学发展史上的一项重大技术突破。

汪建老师一直在推动结直肠癌的大规模筛查，这将是微生物组学在疾病应用方面的一大突破。

无创产前诊断技术（NIPT）成功解决了出生缺陷问题，而健康老龄化的关键在于主要疾病的防控，体质和脑力都希望保持。

衷心感谢爱思唯尔出版集团的 Stacy、Mica、Kavitha、Mohanraj、Selva，以及可能还有更多我不认识的大家。

感谢我的同事们，感谢他们多年来杰出的研究工作。

尤其感谢陈晨、郝梨岚、仝欣、覃友文、鞠艳梅、朱杰、孟彦铮、梁卫婷、揭著业，感谢他们在本书出版过程中在校对文本，提供高分辨率附图等方面做出的贡献。

感谢我亲爱的丈夫，感谢他的支持与鼓励。

感谢关注本书的大家。这次经历对我来说是一个很好的学习机会，我在撰写的过程中，将掌握的各种零散研究信息梳理整合成了更加系统、连贯的图景。关于共生微生物的研究已经呈井喷式爆发，关于共生微生物的研究已经井喷，很抱歉我不能引用每一项新成果。是时候超越单一的粪便样本和粗略的问卷调查，正式进入开展身体"无菌"部位微生物研究的新阶段了。真心期待以后会有更好的研究范例被纳入本书的后续版本，为临床实践以及新的研究与教育提供新的启发。